Advances in Game-Based Learning

Series editor:

Dirk Ifenthaler
Scott JosephWarren
Deniz Eseryel

More information about this series at http://www.springer.com/series/13094

Pieter Wouters • Herre van Oostendorp
Editors

Instructional Techniques to Facilitate Learning and Motivation of Serious Games

Springer

Editors
Pieter Wouters
Utrecht University
Utretcht, The Netherlands

Herre van Oostendorp
Utrecht University
Utrecht, The Netherlands

Advances in Game-Based Learning
ISBN 978-3-319-39296-7 ISBN 978-3-319-39298-1 (eBook)
DOI 10.1007/978-3-319-39298-1

Library of Congress Control Number: 2016950587

Printed on acid-free paper

This Springer imprint is published by Springer Nature
The registered company is Springer International Publishing AG Switzerland

Contents

Chapter 1
Overview of Instructional Techniques to Facilitate Learning and Motivation of Serious Games

Pieter Wouters and Herre van Oostendorp

Abstract Computer games that are used for the purpose of learning, training, and instruction are often referred to as serious games. The last decade shows a huge increase in empirical studies investigating the learning effectiveness and motivational appeal of serious games. Recent meta-analyses show that serious games are effective compared to traditional instruction but that the effectiveness can be improved. This chapter explores which specific instructional techniques can further improve learning and increase motivation. We define instructional techniques as any adaptation of a feature of the game itself or in the context of the game that influences the selection of relevant information, the organization, and integration of that information and/or the intrinsic motivation of the player. The starting point is a meta-analysis conducted in 2013 that is updated and extended. The meta-analysis has a value-added approach and shows which game features can improve learning and/or increase motivation. The interpretation of the results will yield nine proven effective or promising instructional techniques in terms of learning and/or motivation. This set of nine techniques—content integration, context integration, assessment and adaptivity, level of realism, narration-based techniques, feedback, self-explanation and reflection, collaboration and competition, and modeling—form the basis of this volume, which is closed by a reflection chapter.

Keywords Instructional techniques • Serious games • Learning • Motivation • Meta-analysis

1.1 Introduction

In the last decade, we have seen a boost in empirical studies regarding the effectiveness of computer games in learning, training, and instruction. In the literature, such learning environments are often referred to as serious games or game-based

P. Wouters • H. van Oostendorp (✉)
Utrecht University, Utrecht 3512 JE, The Netherlands
e-mail: h.vanOostendorp@uu.nl

P. Wouters, H. van Oostendorp (eds.), *Instructional Techniques to Facilitate Learning and Motivation of Serious Games*, Advances in Game-Based Learning, DOI 10.1007/978-3-319-39298-1_1

learning. With the increasing number of studies also several quantitative and qualitative meta-reviews have been published that have shown the potential of serious games (Boyle et al., 2016; Clark, Tanner-Smith, & Killingsworth, 2015; Ke, 2009; O'Neil, Wainess, & Baker, 2005; Sitzmann, 2011; Vogel et al., 2006; Wouters, van der Spek, & van Oostendorp, 2009; Wouters, Van Nimwegen, Van Oostendorp, & Van Der Spek, 2013; Wouters & Van Oostendorp, 2013). In addition, these reviews have emphasized the conditions under which serious games are effective.

Typically, empirical studies in game research have a particular approach. Mayer (2011, 2016) has divided game research into three categories: a value-added approach with the underlying question how specific game features foster learning and motivation; a cognitive consequences approach, which investigates what people learn from serious games and a media comparison approach, which investigates whether people learn better from serious games than from conventional media. Recent meta-analyses with a media comparison approach reveal that serious games are more effective than traditional learning methods, but that the effect size is only low to moderate. For instance, Wouters et al. (2013) found an effect size of $d=.29$ when comparing serious games with traditional instruction which is in line with the range of effect sizes ($d=.28$ to $d=.37$) that Sitzmann (2011) found when comparing simulation games with non-simulation game groups. Reviews from a value-added approach such as our own (Wouters & Van Oostendorp, 2013) show that the effectiveness of serious games can be improved when specific features are implemented (e.g., we found a strong effect for providing educational feedback) while other features seem to have no or limited effect (e.g., providing advice does not seem to improve the learning potential of a serious game). In other words, it is useful to explore whether the effectiveness of serious games can be increased by implementing or further elaborating specific features of the game.

Theoretically, games may influence learning in two ways, directly by changing the cognitive processes and indirectly by affecting the motivation. Figure 1.1 presents the cognitive–affective model of learning with media in which both dimensions are integrated (Moreno & Mayer, 2007).

From a cognitive perspective, two structures are regarded crucial for the processing of information. First, working memory has a limited capacity to process information and is often not sufficient for learning material that is complex, multimodal, and/or dynamical. Especially for novices, the complexity and the dynamic character of instructional material may lead to problems: they do not know what is relevant and therefore focus on the wrong information. The second structure, long-term memory, has a virtually unlimited capacity, which can serve as added processing capacity by means of schemas, i.e., cognitive structures that can be processed in working memory as a single entity (Kintsch, 1998; Paas, Renkl, & Sweller, 2003). Instructional techniques can support the learner or the player in order to overcome the limitations of the human cognitive architecture (Mayer, 2001, 2011; Paas et al., 2003; Van Oostendorp, Beijersbergen, & Solaimani, 2008).

Based on this cognitive architecture, theories have emphasized several important cognitive processes that are involved in learning. The cognitive–affective model of learning with media (Moreno & Mayer, 2007), for example, discerns three types of

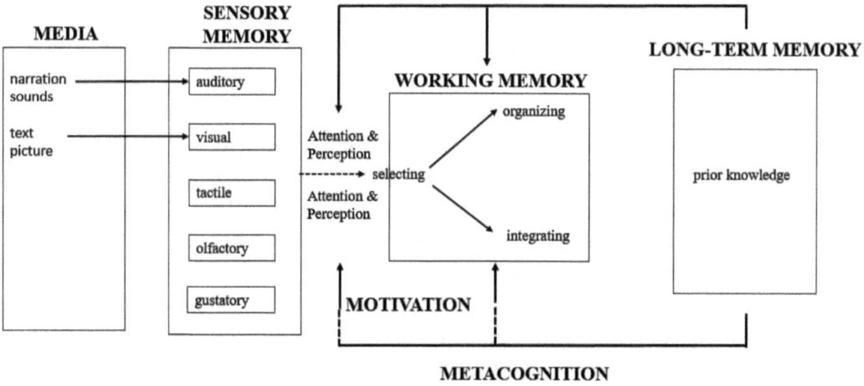

Fig. 1.1 The cognitive–affective model of learning with media

cognitive processing in working memory: *selecting* relevant information by paying attention to the presented material, mentally *organizing* the new information in a coherent structure and *integrating* this structure with prior knowledge (Moreno & Mayer, 2007).

As depicted in Fig. 1.1 auditory sensations such as words, sounds, and music are presented to the ear and visual sensations such as images are presented to the eye through sensory memory. The learner pays attention to specific auditory and visual sensations and organizes the selected information in respectively a verbal and pictorial model. As can be seen in Fig. 1.1, integration involves the merging of the verbal, the pictorial model, and the relevant prior knowledge into one coherent structure. The selection, organization, and integration of information are guided by two mechanisms. To start with, prior knowledge at least partially guides these cognitive processes (as illustrated by the top–down arrows from long-term memory to attention, perception, and working memory). Second, as indicated by the bottom–up arrows from working memory to long-term memory, these cognitive processes are regulated by the metacognitive skills and the motivation.

Although organizing and integrating information reflect different cognitive processes, they are closely related and difficult to separate (cf. Moreno & Mayer, 2005). Therefore, we propose that effective serious games should enable learners to engage in two types of cognitive processes: (1) the selection of relevant information from the learning material and (2) the active organization of that information and the integration with prior knowledge (Mayer & Moreno, 2003).

Motivation influences learning in a more indirect way. Several theories emphasize the potential of serious games to positively influence intrinsic motivation (Garris, Ahlers, & Driskell, 2002; Malone, 1981). This means that players are willing to invest more time and energy in game play not because of extrinsic rewards, but because the game play in itself is rewarding. There is evidence that intrinsic motivation leads to strategies that demand more effort and enable a deeper level of processing (Lumsden, 1994). In addition, intrinsically motivated children use more logical information-gathering and decision-making strategies, and prefer activities of higher difficulty in comparison with extrinsically motivated children (Deci, 1975; Lepper, 1988). Several characteristics of serious games have been identified to explain this motivating appeal. Malone (1981) proposed that the most important factors that make playing a computer game intrinsically motivating are challenge, curiosity, and fantasy. Two other essential factors associated with motivation and computer games, autonomy (i.e., the opportunity to make choices) and competence (i.e., a task is experienced as challenging but not too difficult) originate from self-determination theory and are known to positively influence the experienced motivation (Przybylski, Rigby, & Ryan, 2010).

Serious games can be complex (learning) environments, and it is not always the case that students playing serious games will automatically engage in the aforementioned cognitive processes or that the characteristics of the game trigger (intrinsic) motivation. For example, players can be easily overwhelmed by the plentitude of information, the multimodal presentation of information (sometimes simultaneously on different locations of the screen), the choices players potentially can make, the dynamics of the game and the complexity of the task that has to be performed. These demands on the cognitive resources can make it difficult for novices to discern between relevant and irrelevant information and select the information that is required. Novices can also become easily demotivated when they are overwhelmed by information and options while they do not know what to do. In addition, in computer games players act and see the outcome of their actions reflected in changes in the game world. This may lead to a kind of intuitive learning: players know how to apply knowledge, but they cannot explicate it (Leemkuil & de Jong, 2011). Yet, it is important that learners articulate and explain their knowledge because it urges them to *organize* new information and *integrate* it with their prior knowledge. Ultimately, this will yield a knowledge base with higher accessibility, better retention, and higher transfer of learning (Wouters, Paas, & van Merriënboer, 2008).

The focus of this book is to improve the effectiveness of serious games through the application of instructional techniques that address the required cognitive processes and/or increase intrinsic motivation. We define instructional techniques as any adaptation of a feature of the game itself or in the context of the game that influences the selection of relevant information, the organization and integration of that information, and/or the intrinsic motivation of the player. In this respect, an instructional technique may apply to a relatively small manipulation such as a prompt that triggers a player to explain a specific answer as well as complementing the serious game with other instruction methods such as a class discussion. As starting point, we use the results of a meta-analysis that we conducted earlier in which we investigated the effect of instructional support in serious games on learning (Wouters & Van Oostendorp, 2013). This meta-analysis revealed a moderate effect size, indicating that instructional support overall yields learning compared to serious games without such support. However, the results also showed that not every type of instructional support is beneficial and that types of support that address the selection of relevant information are more effective than support that helps to organize/integrate new information. Since the publication of the meta-analysis, many empirical studies have been published. This is relevant because some types of support were reported only once in the original meta-analysis and an update will possibly reveal new studies of these types of support so that they can represent an instructional technique on itself (e.g., this was the case for adaptivity). In this chapter, we will update this meta-analysis, extend it with new instructional techniques and present a comprehensive set of proven effective or promising instructional techniques that will be further developed/elaborated in the subsequent chapters.

1.2 Serious Games

Several scholars have provided definitions or classifications of computer games characteristics (Garris et al., 2002; Malone, 1981; Prensky, 2001). For the purpose of this meta-analysis, we describe computer games in terms of being *interactive* (Prensky, 2001; Vogel et al., 2006), based on a set of *agreed rules and constraints* (Garris et al., 2002), and directed towards a clear *goal* that is often set by a *challenge* (Malone, 1981). In addition, games constantly provide *feedback* either as a score or as changes in the game world to enable players to monitor their progress towards the goal (Prensky, 2001). Some scholars contend that computer games also involve a competitive activity (against the computer, another player, or oneself), but it can be questioned if this is a defining characteristic. Of course, there are many games in which the player is in competition with another player or with the computer, but in a game like SimCity players may actually enjoy the creation of a prosperous city that satisfies their beliefs or ideas without having the notion that they engage in a competitive activity. In the same vein, a narrative or the development of a story can be very important in a computer game (e.g., in adventure games), but again it is not a prerequisite for being a computer game (e.g., action games do not really require a narrative). This definition of GBL would also comprise "pure simulations" such as SIMQuest (see also www.simquest.nl) that also include an underlying model in which learners can provide input (either changing variables or perform actions) and observe the consequences of their actions (cf. Leemkuil, de Jong, & Ootes, 2000). However, we concur with Jacobs and Dempsey (1993) who argued that task-irrelevant elements are often removed from simulations, whereas other elements such as an engaging context are included or emphasized to define a (simulation) game. In GBL, the objective of the computer game is not to entertain the player, which would be an added value, but to use the entertaining quality for training, education, health, public policy, and strategic communication objectives (Zyda, 2005).

1.3 Instructional Techniques

Table 1.1 provides an overview of instructional techniques that we found in the original meta-analysis, examples, the cognitive process(es) that they are—arguably—associated with, and whether they explicitly aim to influence the motivation of the player.

Two instructional techniques need a clearer explanation because they cannot directly be derived from the meta-analysis. As far as we know, these techniques are not yet investigated from a value-added approach. Yet, we regard them as adaptations in the game or part of the context of the game that can improve learning or increase the motivation.

As mentioned before, a criticism on serious games is that the acquired knowledge and skills remain implicit, and therefore are difficult to apply in new situations. Table 1.1 discerns several instructional techniques that enable students to explicate the newly acquired knowledge and skills (e.g., prompting self-explanations). However, these instructional techniques are implemented within the game, whereas it is also possible to use other instruction methods to activate prior knowledge or to reflect explicitly on the new knowledge and skills (e.g., a class discussion). In this book, we use the term Context integration to refer to such an instructional technique. Several media-comparison reviews clearly show that serious games combined with other instruction methods are potentially more effective than serious games in isolation (cf. Sitzmann, 2011; Wouters et al., 2013; Young et al., 2012).

The second instructional technique, Level of Realism, is interesting for several reasons. To start with, it is known from multimedia research that spoken explanations of visual information are more effective than written explanations, especially when the visual channel becomes overloaded (Mayer, 2011). The question that can be raised is whether this is true for serious games as well. Secondly, Level of Realism involves several dimensions. In Table 1.1, we have outlined these dimensions. Besides modality, it also includes the visual representation within a game and multimodal interactions. As far as we know, there are no studies that have examined how different visual representations affect learning and motivation, but there are

Table 1.1 Overview of instructional techniques, examples of the technique, the associated cognitive processes, and the motivational characteristic

Instructional technique	Examples	Cognitive processes	Motivation
Adaptivity/Assessment Adapting complexity/difficulty of game tasks to the abilities of the student through real-time assessment	Adaptivity	Selection Organization/ Integration	No
Advice System generated information to support the learner to continue in the game (e.g., by focusing attention)	All types of advice whether contextualized, adaptive, or not	Selection	No
Collaboration Working in groups with discussion, often aiming at the explication of implicit knowledge	Players played in dyads, groups, or engaged in group discussion	Organization/ Integration	No
Content Integration Learning content is integrated with game mechanics	Intrinsic integration	Selection Organization/ Integration	Yes

(continued)

Table 1.1 (continued)

Instructional technique	Examples	Cognitive processes	Motivation
Context Integration The combination of a serious game with other instruction methods (e.g., a class discussion)		Selection Organization/ Integration	No
Feedback Information is given whether an answer or action is correct or not. The feedback can be corrective (correct or not), explanatory (why correct or not)	Feedback, Guidance	Selection	No
Interactivity Learners make choices in the game in order to solve a problem or to perform a task	Interactivity, learner control, and choice of game features	Organization/ Integration	Yes
Level of Realism The use of both the auditory channel (e.g., spoken text, sounds, music) and the visual channel. Also the type of auditory and visual representation	Modality, Sounds, Music, Visual design	Selection	Yes
Modeling An explication or indication how to solve a problem. The explanation can be given by a peer or expert and can be verbal, animated, or graphical	Different types of scaffolding, modeling, worked examples	Selection Organization/ Integration	No
Narrative elements A narrative context or manipulation of narrative elements which provide a cognitive framework	Fantasy, rich narrative, foreshadowing, surprising events, curiosity	Selection Organization/ Integration	Yes
Personalization Ideas, characters, topics, and messages are presented in such a way that they have a specific high interest value for the learner/player	Personalization, Personalized messages	Unknown	Yes
Reflection Learners are stimulated to think about their answers and (sometimes) explain it to themselves	Reflection, self-explanation, elaboration, collaboration, worked example	Organization/ Integration	No

studies that investigate how different visual representations affect the perception of the game and the quality of task performance in the game (e.g., Rojas, Kapralos, Collins, & Dubrowski, 2014). In addition, a media comparison meta-analysis that we conducted to compare serious games with traditional instruction showed that the better performance of serious games relative to traditional instruction was particularly true for more basic visual representations (Wouters et al., 2013). These findings justify a closer examination of the role of visual representations. Finally, computer games provide an opportunity for multiple modalities, and the question can be raised how multimodal interactions (e.g., sounds and visual representations) have an impact on learning and motivation.

1.4 Method

In order to update, the meta-analysis we conducted an additional search to find new studies. The most important criteria for selection were that studies compared relevant groups (e.g., a group with a type of instructional support and a group without this support) and that sufficient data had to be available to calculate an effect size (we used Cohen's *d*). The following sources were used:

1. Search actions in Google Scholar. The search terms we used were: "Game-based learning," "serious games," "educational games," "simulation games," "virtual environments," and "muve." When necessary, these search terms were combined with "learning," "instruction," and "training."
2. References in meta-analyses/reviews that were published since our own meta-analysis (2013).
3. We also inspected which studies referred to meta-analyses/reviews (through the citations option in Google Scholar.

For the new studies, we coded: the type of instructional technique, types of cognitive processes that were involved, the visual design, and whether motivation was measured. Cohen's *d* was used as effect size measure. This effect size was corrected in the following situations:

- When the study used a small sample
- When a condition in a study was used for multiple comparisons
- When multiple constructs were used for learning outcome or (perceived) motivation
- When one construct for learning outcome or (perceived) motivation was measured in multiple ways

Given the great variation in types of serious games and features, it is likely that the average effect sizes in the populations vary between the studies. Therefore, we used the random-effects model for the main analyses and the moderator analyses with 95 % confidence intervals around the weighted mean effect sizes. For a more extensive explanation of the procedure, we refer to Wouters and Van Oostendorp (2013).

1.5 Results

For *learning*, the search for new studies revealed 21 studies with 39 pairwise comparisons (some studies allow multiple comparisons, for example, because one control group was compared with two different instructional techniques). This updated meta-analysis therefore involved 50 studies (original meta-analysis: 29 studies) with 146 pairwise comparisons (original meta-analysis: 107 pairwise comparisons). See the Appendix New Studies for the 21 new studies. The weighted mean effect size of learning for instructional techniques was $d=.41$ ($z=7.29, p<.001$) indicating that in general serious games with instructional techniques improve learning more than serious games without instructional techniques.

Since *motivation* was not included in the original meta-analysis, not only a new search was conducted but also the motivation data of the studies of the original meta-analysis were included. For motivation, the updated meta-analysis involved in total 17 studies with 62 pairwise comparisons. With regard to motivation, serious games with instructional techniques are in general more motivating than those without instructional techniques ($d=.26$, $z=4.17$, $p<.001$). Table 1.2 shows the effect sizes for learning and motivation for each instructional technique. The table also includes two moderator analyses: one for cognitive processes and one for visual representation. We regard the latter as an extension of Level of Realism.

With respect to learning the instructional techniques Adaptivity/Assessment, Collaboration, Feedback, Modeling, Level of Realism, Personalization, and Reflection improve learning significantly relative to serious games without these techniques. This pattern is similar to the results of the earlier meta-analysis. A qualification on type of cognitive processes shows that instructional techniques defined as game feature are most effective when they address the selection of relevant information ($d=.60$) and less when they only aim at organizing and integrating new information ($d=.16$) (see Table 1.2). The moderator analysis on Visual Representation shows that instructional techniques in serious games with basic and cartoon-like visual representations are more effective than instructional techniques in serious games with a (photo) realistic design. In fact, instructional techniques do not seem to be effective when applied in (photo) realistic designs.

Although results indicate that in general instructional techniques have a positive effect on the perceived motivation (overall mean $d=.26$), a closer look reveals a dependency on the type of instructional technique that is used. Especially instructional techniques such as Level of Realism, Narrative elements, and Personalization yield a higher level of perceived motivation relative to serious games without these techniques, whereas others seem to have no effect.

1.6 Conclusions

The goal of this chapter was to gain insight into the impact of several instructional techniques on learning and motivation. Moreover, we have mapped the instructional techniques to one or more cognitive processes that are relevant for learning: selecting, organizing, and integrating information with prior knowledge. Instructional technique was defined as any adaptation of a feature of the game itself or in the context of the game that influences the selection of relevant information, the organization and integration of that information, and/or the intrinsic

Table 1.2 Effect sizes for learning and motivation for each instructional technique and the moderator analysis for cognitive processes and visual representation

	Learning			Motivation		
	d	*k*	95 % CI	*d*	*k*	95 % CI
Instructional technique						
Adaptivity/Assessment	1.84	3	[−1.24, 5.45]	.77	4	[−7.76, 9.30]
Advice	.12	16	[−.11, .35]	–	–	–
Collaboration	.16*	18	[.02, .46]	−.02	2	[−.74, .70]
Content integration	.26	6	[−.21, .78]	1.68	2	[−14.21, 17.56]
Context integration	na			na		
Feedback	.63***	8	[.33, 1.87]	.41	2	[-.04, .86]
Interactivity	.13	12	[-.13, .39]	.25	11	[.25, .74]
Level of Realism	1.24***	10	[.90, 3.67]	.49***	6	[.25, .74]
Modeling	.55***	13	[.22, 1.63]	.01	4	[−.29, .31]
Narrative elements	.10	9	[−.12, .29]	.35*	8	[.07, .63]
Personalization	1.06***	4	[.50, 3.13]	1.18*	4	[.23, 2.12]
Reflection	.23**	20	[.08, .67]	−.05	6	[−.34, .25]
Other	na	27		na	17	
Cognitive processes						
Selection	.60***	46	[.42, 1.78]	.22	12	[-.98, 1.42]
Organization/integration	.16***	48	[.07, .47]	.20*	20	[.01, .40]
All processes	.55	16	[−.03, 1.63]	.92	9	[−3.88, 5.73]
Visual representation						
Schematic	.41***	80	[.30. 1.23]	.35***	35	[.18, .52]
Cartoon-like	.39*	28	[.03, 1.16]	.20	12	[−.09, .48]
Realistic	.08	25	[−.05, .24]	.17	16	[−.05, .38]
Mixed	.60	13	[.17, 1.77]	na		

Note: *$p<.05$, **$p<.005$, ***$p<.001$
d=effect size measure, k=number of pairwise comparisons, CI=confidence interval for d. Effect sizes for cognitive processes only mentioned for the instructional techniques (Other is left out). Level of Realism includes only modality studies. As far as we know, no value-added Context Integration studies have been conducted

motivation of the player. The starting point was a meta-analysis from a value-added perspective performed by Wouters and Van Oostendorp (2013) which was updated with new studies. The updated meta-analysis confirmed that the effectiveness can be improved and the motivation increased by implementing specific instructional techniques.

With respect to the relevant learning processes, particularly instructional techniques supporting the selection of relevant information improve the learning potential of serious games. Especially *Modeling* (showing which information is important in order to solve a problem and how to solve a problem) and *Feedback* (information whether and/or why an answer is correct) are effective techniques to support learners in selecting relevant information. It seems more difficult to implement instructional techniques in such a way that learners are prompted to actively engage in the organization and integration of new information. The only exception is *Reflection* in which learners are explicitly asked to think about their actions or the answer they have given and in this way are stimulated to make intuitive knowledge more explicit.

Although *Personalization* has a strong effect on both learning and motivation, the problem arises that it is based on four comparisons from only one study. *Narrative elements*—the application of techniques from text and discourse processing such as curiosity, surprise, suspense, fantasy—have no direct effect on learning, but they do have a positive effect on motivation. In the updated meta-analysis, a robust but low effect was found for *Collaboration*. A technological challenge in serious game design is the implementation of *Adaptivity/Assessment*. One of the problems is the implementation of real-time assessment of game performance in such a way that it can be used to adapt the difficulty of the game to an appropriate level. Although the technical challenge is reflected in the small number of pairwise comparisons (three comparisons from three studies), the results also show that this technique potentially can have a very strong learning effect. These considerations, however, make a closer investigation of *Adaptivity/Assessment* valuable.

Although *Content Integration* is propagated in the literature, the results of the meta-analysis give an ambiguous picture for both learning and motivation. It seems to depend on the specific implementation of the content integration whether it is very effective and motivating or not. This suggests that content integration is potentially a promising instruction technique but that it is not yet clear how the domain content and game mechanics can best be integrated. These considerations, however, make a closer investigation of content integration valuable.

With regard to *Level of Realism*, the updated meta-analysis confirms that the multimedia modality effect (spoken explanations accompanying visual information are more effective than written explanations, Mayer, 2011) is also found in serious games. Although we found no studies in which the effect of two or more visual representations on learning and motivation was investigated, a moderator analysis shows that instructional techniques in serious games with basic and cartoon-like visual representations are more effective than instructional techniques in serious games with a (photo)-realistic design. These findings are corroborated by other reviews (Clark et al., 2015; Wouters et al., 2013). A possible explanation is that schematic/cartoon-like designs facilitate students to focus on relevant information in the game, whereas in games with (photo)-realistic designs students can easily become overwhelmed by the visual complexity.

Evidence for *Context Integration* comes from the reviews by Sitzmann (2011) and Young et al. (2012). Normally, students gain only intuitive knowledge during game play and are not prompted to verbalize the new knowledge and in this way don't anchor it more profound in their knowledge base (Leemkuil & de Jong, 2011; Wouters et al., 2008). The beneficial effect of supplemental instructional methods is that they prompt or support players to articulate the new knowledge and integrate it with their prior knowledge. Altogether, the conclusion of the results of the updated meta-analysis is that the implementation of instructional techniques in general

Table 1.3 List of effective or promising instructional techniques

Instructional technique	Learning	Motivation
Adaptivity/Assessment	×	
Collaboration	×	
Content integration	×	×
Context integration	×	
Feedback	×	
Level of Realism	×	×
Modeling	×	
Narrative elements		×
Reflection	×	

improves learning and potentially increases motivation. However, the results show that not all relevant cognitive processes are equally addressed and not all instructional techniques are equally effective. Table 1.3 lists instructional techniques that have proven to be effective or for which there are indications that they are promising.

The results of the updated meta-analysis also show that it is not clear under what conditions these instructional techniques are effective, which (technical) challenges they face in order to become (more) effective and what guidelines serious games designers should consider when they choose to implement an instructional technique. In the remainder of this book, experts will further elaborate on these instructional techniques.

1.7 Some Limitations Regarding This Analysis

A meta-analysis aggregates the results of studies that share a common characteristic, but these studies are also different on many other characteristics (the type of game, the domain, the age of the participants, etc.). All these characteristics may have an impact on the effect of a specific instructional technique on learning and motivation. In the subsequent chapters, the contributors will address these issues.

Some of the characteristics of this meta-analysis should be kept in mind when interpreting the results. To start with, we used a broad definition of instructional technique which may imply that it is difficult to generate and interpret the weighted mean effect size. We have tried to overcome this by adopting a random-effects model and a further qualification of the effect of instructional technique on the cognitive processes that they aim at (selection vs. organization/integration). In some cases, this may be arbitrary. For example, we have classified "feedback" as a type of instructional technique that draws the focus of a player's attention to relevant information, whereas it is also plausible that it triggers players to engage in organizing and integrative activities. In these cases, we are guided by our assessment of the most important cognitive process as described in the study, but we realize that other valid choices are possible as well (see also Ke, 2016).

This review considers the moderator variables separately. For example, we investigated the effect of different types of instructional technique on learning in general, but we do not know how these different types of instructional technique influence the different types of learning outcome such as knowledge, skills, and in-game performance. This means that we have no understanding of the interactions between the moderator variables.

Most studies measure learning immediately or shortly (ranging from some days to a week) after playing the game. In our media comparison meta-analysis, we found that the learning effect is more persistent in serious games than in traditional instruction methods. This "retention effect" is important because it supports what teachers and instructors deem important: that serious games lead to well-structured prior knowledge on which learners can build on during their learning career. Researchers should also put more effort in investigating the long-term effects of the different instructional techniques and unravel under what conditions learning is persistent.

Perhaps the foremost reason to use serious games is their alleged motivational appeal (Garris et al., 2002; Malone, 1981). However, this assumption should be treated with caution. The motivational appeal of games is often attributed to the fact that players experience a sense of control regarding their actions and decisions in the game. The other side of the coin is that there is no control or only limited control of the game in the instructional context: often students have to play a predetermined game during specific class hours (see also the comments of Graesser in Chap. 11 of this volume). Another critical comment can be made regarding the measurement. In most cases, motivation is measured with surveys that are administered long after the motivational drive arises. It is known that the experienced motivational drive to act is present at the moment that there exists a need, after the need is fulfilled the intensity of motivation decreases (Maslow, Frager, Fadiman, McReynolds, & Cox, 1970).

1.8 Organization of This Volume

In the remainder of the volume, the selected instructional techniques are discussed in more detail. In Chap. 2 *Towards a game-based learning instructional design model focusing on integration* (Vandercruysse) and Elen position Content integration in their instructional model of game-based learning. Martinez-Garza and Clark present in Chap. 3 *Two systems, two stances: a novel theoretical framework for model-based learning in digital games* a novel theoretical framework for model-based learning in serious games and its implications for the design and use of games in a broader context. Chapter 4 *Assessment and adaptation in games* by Shute, Ke, and Wang describes in detail stealth assessment during game play and how it can be used to create adaptivity. In Chap. 5, *Fidelity and multimodal interactions*, Kapralos, Moussa, Collins, and Dubrowski investigate how multimodal (audio–visual) interactions have an impact on fidelity, performance, and computational requirements. Van Oostendorp and Wouters discuss the role of narrative elements such as curiosity and surprise in Chap. 6 *Narration-based techniques to facilitate game-based learning*. In Chap. 7, *Designing effective feedback messages in serious games and simulations: a research review*, Johnson, Bailey, and Van Buskirk give an overview of several dimensions of feedback in games and how they influence learning. In Chap. 8, *Self-explanations in game-based learning: from tacit to transferable knowledge*, terVrugte and de Jong present self-explanations, and methods to implement them in serious games, as a way to foster reflection in serious games. Chapter 9 *Competition and collaboration for game-based learning: a case study* by Sanchez discusses how competitive systems, that is, the combination of collaboration and competition influences learning (processes) and motivation. Wouters describes in Chap. 10 *Modeling and worked examples in game-based learning* how modeling and worked examples can be classified on four dimensions and how they have an impact on both learning and motivation. In the final chapter, *Reflections on serious games*, Graesser provides some comments on the contributions in the book volume and presents some challenges for the (near) future.

References

Boyle, E. A., Hainey, T., Connolly, T. M., Gray, G., Earp, J., Ott, M., et al. (2016). An update to the systematic literature review of empirical evidence of the impacts and outcomes of computer games and serious games. *Computers & Education, 94*, 178–192.

Clark, D. B., Tanner-Smith, E. E., & Killingsworth, S. S. (2015). Digital games, design, and learning a systematic review and meta-analysis. *Review of Educational Research*, 0034654315582065.

Deci, E. L. (1975). *Intrinsic motivation*. New York: Plenum.

Garris, R., Ahlers, R., & Driskell, J. E. (2002). Games, motivation, and learning: A research and practice model. *Simulation and Gaming, 33*, 441–467.

Jacobs, J. W., & Dempsey, J. V. (1993). Simulation and gaming: Fidelity, feedback and motivation. In J. V. Dempsey & G. C. Sales (Eds.), *Interactive instruction and feedback*. Englewood Cliffs, NJ: Educational Technology Publications.

Ke, F. (2009). A qualitative meta-analysis of computer games as learning tools. In R. E. Ferdig (Ed.), *Handbook of research on effective electronic gaming in education* (Vol. 1, pp. 1–32). Hershey, PA: Information Science Reference.

Ke, F. (2016). Designing and integrating purposeful learning in game play: A systematic review. *Educational Technology Research and Development, 64*, 219–244.

Kintsch, W. (1998). *Comprehension: A paradigm for cognition*. Cambridge, MA: Cambridge University Press.

Leemkuil, H., & de Jong, T. (2011). Instructional support in games. In S. Tobias & D. Fletcher (Eds.), *Computer games and instruction* (pp. 353–369). Charlotte, NC: Information Age Publishing.

Leemkuil, H., de Jong, T., & Ootes, S. (2000). *Review of educational use of games and simulations* (IST-1999-13078 Deliverable D1).University of Twente, Enschede, The Netherlands.

Lepper, M. R. (1988). Motivational considerations in the study of instruction. *Cognition and Instruction, 5*(4), 289–309.

Lumsden, L. S. (1994). Student motivation to learn. *ERIC Digest*, No. 92, pp. 1–3.

Malone, T. (1981). Toward a theory of intrinsically motivating instruction. *Cognitive Science, 4*, 333–369.

Maslow, A. H., Frager, R., Fadiman, J., McReynolds, C., & Cox, R. (1970). *Motivation and personality* (Vol. 2). New York: Harper & Row.

Mayer, R. E. (2001). *Multimedia learning*. New York: Cambridge University Press.

Mayer, R. E. (2011). Multimedia learning and games. In S. Tobias & J. D. Fletcher (Eds.), *Computer games and instruction* (pp. 281–305). Charlotte, NC: Information Age Publishing.

Mayer, R. E. (2016). What should be the role of computer games in education? *Policy Insights from Behavioral and Brain Sciences*, 3(1), 20–26.

Mayer, R. E., & Moreno, R. (2003). Nine ways to reduce cognitive load in multimedia learning. *Educational Psychologist, 38*, 43–52.

Moreno, R., & Mayer, R. E. (2005). Role of guidance, reflection, and interactivity in an agent-based multimedia game. *Journal of Educational Psychology, 97*, 117–128.

Moreno, R., & Mayer, R. (2007). Interactive multimodal learning environments. *Educational Psychology Review, 19*(3), 309–326.

O'Neil, H. F., Wainess, R., & Baker, E. L. (2005). Classification of learning outcomes: Evidence from the computer games literature. *The Curriculum Journal, 16*, 455–474.

Paas, F., Renkl, A., & Sweller, J. (2003). Cognitive load theory and instructional design: Recent developments. *Educational Psychologist, 38*(1), 1–4.

Prensky, M. (2001). *Digital game-based learning*. New York: McGraw-Hill.

Przybylski, A. K., Rigby, C. S., & Ryan, R. M. (2010). A motivational model of video game engagement. *Review of General Psychology, 14*, 154–166.

Rojas, D., Kapralos, B., Collins, K., & Dubrowski, A. (2014). The effect of contextual sound cues on visual fidelity perception. *Studies in Health Technology and Informatics, 196*, 346–352.

Sitzmann, T. (2011). A meta-analytic examination of the instructional effectiveness of computer-based simulation games. *Personnel Psychology, 64*, 489–528.

Van Oostendorp, H., Beijersbergen, M. J., & Solaimani, S. (2008, June). Conditions for learning from animations. In *Proceedings of the 8th international conference on International conference for the learning sciences* (Vol. 2, pp. 438–445). International Society of the Learning Sciences.

Vogel, J. J., Vogel, D. S., Cannon-Bowers, J., Bowers, C. A., Muse, K., & Wright, M. (2006). Computer gaming and interactive simulations for learning: A meta-analysis. *Journal of Educational Computing Research, 34*, 229–243.

Wouters, P., Paas, F., & van Merriënboer, J. J. M. (2008). How to optimize learning from animated models: A review of guidelines based on cognitive load. *Review of Educational Research, 78*, 645–675.

Wouters, P., van der Spek, E. D., & van Oostendorp, H. (2009). Current practices in serious game research: A review from a learning outcomes perspective. In T. M. Connolly, M. Stansfield, & L. Boyle (Eds.), *Games-based learning advancements for multisensory human computer interfaces: Techniques and effective practices* (pp. 232–255). Hershey, PA: IGI Global.

Wouters, P., Van Nimwegen, C., Van Oostendorp, H., & Van Der Spek, E. D. (2013). A meta-analysis of the cognitive and motivational effects of serious games. *Journal of Educational Psychology, 105*(2), 249.

Wouters, P., & Van Oostendorp, H. (2013). A meta-analytic review of the role of instructional support in game-based learning. *Computers & Education, 60*(1), 412–425.

Young, M. F., Slota, S., Cutter, A. B., Jalette, G., Mullin, G., Lai, B., et al. (2012). Our princess is in another castle a review of trends in serious gaming for education. *Review of Educational Research, 82*(1), 61–89.

Zyda, M. (2005). From visual simulation to virtual reality to games. *Computer, 38*(9), 25–32.

Appendix New Studies

Barzilai, S., & Blau, I. (2014). Scaffolding game-based learning: Impact on learning achievements, perceived learning, and game experiences. *Computers & Education, 70*, 65–79.

DeLeeuw, K. E., & Mayer, R. E. (2011). Cognitive consequences of making computer-based learning activities more game-like. *Computers in Human Behavior, 27*, 2011–2016.
Erhel, S., & Jamet, E. (2013). Digital game-based learning: Impact of instruction and feedback on motivation and learning effectiveness. *Computers & Education, 76*, 156–167.
Fiorella, L., & Mayer, R., E. (2012). Paper-based aids for learning with a computer-based game. *Journal of Educational Psychology, 104*, 1074-1082.
Goodman, D., Bradley, N. L., Paras, B., Williamson, I. J., & Bizzochi, J. (2006). Video gaming promotes concussion knowledge acquisition in youth hockey players. *Journal of Adolescence, 29*, 351–360.
Habgood, M. P. J., & Ainsworth, S. E. (2009). Motivating children to learn effectively: Exploring the value of intrinsic integration in educational games. *Journal of the Learning Sciences, 20*, 169–206.
Hwang, G.-J., Sung, H.-Y., Hung, C.-M., Huang, I., & Tsai, C.-C. (2012). Development of a personalized educational computer game based on students' learning styles. *Educational Technology Research & Development, 60*, 623–638.
Hwang, G.-J., Yang, L.-H., & Wang, S.-Y. (2013). A concept map-embedded educational computer game for improving students' learning performance in natural science courses. *Computers & Education, 69*, 121–130.
Koops, M., & Hoevenaar, M. (2013). Conceptual change during a serious game: Using a lemniscate model to compare strategies in a physics game. *Simulation and Gaming, 44*(4), 544–561.
O'Neil, H. F., Chung, G. K. W. K., Kerr, D., Vendlinski, T. P., Buschang, R. E., & Mayer, R. E. (2014). Adding self-explanations prompts to an educational computer game. *Computers in Human Behavior, 30*, 23–28.
Ozcelik, E., Cagiltay, N. E., & Ozcelik, N. S. (2013). The effect of uncertainty on learning in game-like environments. *Computers & Education, 67*, 12–20.
Plass, J. L., O'Keefe, P. A., Homer, B. D., Case, J., Hayward, E. O., Stein, M., et al. (2013). The impact of individual, competitive, and collaborative mathematics game play on learning, performance, and motivation. *Journal of Educational Psychology, 105*(4), 1050–1066.
Sampayo-Vargas, S., Cope, C. J., He, Z., & Byrne, G. J. (2013). The effectiveness of adaptive difficulty adjustments on students' motivation and learning in an educational computer game. *Computers & Education, 69*, 452–462.
Serge, S. R., Priest, H. A., Durlach, P. J., & Johnson, C. I. (2013). The effects of static and adaptive performance feedback in game-based training. *Computers in Human Behavior, 29*, 1150–1158.
Shen, C.-Y., & O'Neil, H. (2006). *The effectiveness of worked examples in a game-based learning environment*. A Paper Presented at the Annual Meeting of the AERA. San Francisco, CA.
Sung, H.-Y., & Hwang, G.-J. (2012). A collaborative game-based learning approach to improving students' learning performance in science courses. *Computers & Education, 63*, 43–51.
ter Vrugte, J., de Jong, T., Wouters, P., Vandercruysse, S., Elen, J., & Van Oostendorp, H. (2015a). When a game supports prevocational math education but integrated reflection does not. *Journal of Computer Assisted Learning, 31*(5), 462–480.
ter Vrugte, J., de Jong, T., Wouters, P., Vandercruysse, S., Elen, J., & Van Oostendorp, H. (2015b). How heterogeneous collaboration and competition interact in prevocational game-based math education. *Computers & Education, 89*, 42–52.
Vandercruysse, S., ter Vrugte, J., de Jong, T., Wouters, P., Van Oostendorp, H., Verschaffel, L., et al. (2014). *Content integration as a factor in math-game effectiveness.*
Wouters, P., Oostendorp, H., ter Vrugte, J., Vandercruysse, S., de Jong, T. & Elen, J. (in press). The effect of surprising events in a serious game on learning mathematics. *British Journal of Educational Technology.* doi:10.1111/bjet.12458
Wouters, P., van Oostendorp, H., ter Vrugte, J., Vandercruysse, S., de Jong, T., & Elen, J. (2015). The role of curiosity-triggering events in game-based learning for mathematics. In J. Torbeyns, E. Lehtinen, & J. Elen (Eds.), *Describing and studying domain-specific serious games* (pp. 191–208). Heidelberg: Springer.

Chapter 2
Towards a Game-Based Learning Instructional Design Model Focusing on Integration

Sylke Vandercruysse and Jan Elen

Abstract This chapter focuses on a new instructional design model for game-based learning (GBL) that pinpoints the elements that are to be considered when designing learning environments in which GBL occurs. One key element of the model is discussed more in detail, being the integration of instructional elements in a GBLE. Based on different studies, the chapter concludes with emphasizing the importance of the design of the GBLE in the GBL processes. More specifically, the interplay between the instructional elements and the game elements is an important aspect in the GBL-process. Several decisions have to be made when designing a GBLE, and these decisions are of influence on GBL outcomes.

Keywords Instructional design model • Game-based learning • Integration of instructional elements

2.1 Introduction

To support students' development of knowledge and skills, educators are busy with developing learning environments. These learning environments aim at facilitating students' learning processes. Therefore, an optimal design of these environments is warranted. However, designing such learning environments is difficult, since many decisions have to be made based on different learning processes, different knowledge components, different teaching methods, etc. (Aleven, Koedinger, Corbett, & Perfetti, 2015). In order to support educators in this design process, different instructional design models exist. Some examples are the elaboration theory of Reigeluth (Reigeluth, Merrill, Wilson, & Spiller, 1980), Merrills' first principles of instruction (Merrill, 2002), or Gagné's nine events of instruction (Gagné, Briggs, & Wager, 1992).

S. Vandercruysse (✉) • J. Elen
CIP&T, Center for Instructional Psychology and Technology, KU Leuven, Dekenstraat 2, 3000 Leuven, Belgium
e-mail: sylke.vandercruysse@kuleuven.be

P. Wouters, H. van Oostendorp (eds.), *Instructional Techniques to Facilitate Learning and Motivation of Serious Games*, Advances in Game-Based Learning, DOI 10.1007/978-3-319-39298-1_2

In these models, instructional designers try to outline ideal instructional methods for predetermined outcomes with a specific group of learners. Hence, these models give structure and meaning to instructional design problems and questions by breaking them down into discrete, manageable units (Ryder, 2015). The value of these models is determined by the context of use, since for instance, each model assumes a specific intention of its user (Ryder, 2015). This implies that the one "optimal" learning environment does not exist and depends on different aspects, e.g., the learners, the learning goals, and the context. This chapter focuses on a new instructional design model for game-based learning (GBL) that pinpoints the elements that are to be considered when designing learning environments in which GBL occurs. More specifically, one key element will be discussed more in detail, being the integration of instructional elements in a GBLE (see further). In this chapter, GBL refers to learning (outcomes) occurring from learning processes in which learners use an educational game.

Although general instructional design models can be used for GBL, specific models focusing on GBL have also been elaborated. One such GBL model is proposed by Garris, Ahlers, and Driskell (2002; the input-process-outcome model). This model tries to visualize how and when learning occurs when learners play a game. The input represents the educational game consisting of instructional content, mixed with game characteristics. During the game process, the learners are expected to repeat cycles within a game context. The learning outcomes, in turn, are conceptualized as a multidimensional construct of learning skills, cognitive outcomes, and attitudes. Another model of GBL is presented by Liu and Rojewski (2013). This model stresses that, in order to achieve GBL, an appropriate game design is essential, as well as an optimal game application or implementation. This indicates that not only the GBLE matters, but also the way the GBLE is applied or implemented in instructional activities.

Both models show a different elaboration of a learning environment in order to obtain GBL. However, they both emphasize the same basic idea: there is a need for a well-designed educational game environment, as well as for an effective game application. Hence, taking both models together, a new instructional design model for GBL can be constructed (see Fig. 2.1). The aspects that appeared essential in both models are also two central elements of the new model. However, a third key

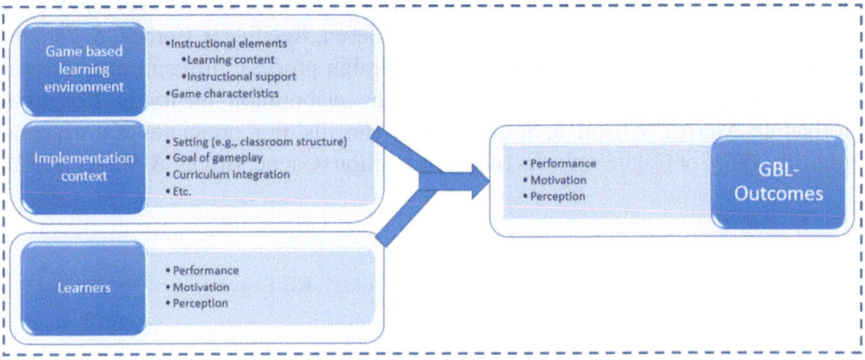

Fig. 2.1 Instructional design model of game-based learning

element is added to the model, being the learners. This element is not completely lacking in the two previous mentioned GBL models, but the learners were only included implicitly. However, because of the importance of this element, in the new GBL model the learners are not only implicitly included, but are incorporated as a third crucial element to be considered while designing for GBL. Finally, the learning outcomes are—in line with the two above-mentioned models—conceptualized as multidimensional: not only cognitive outcomes are assumed, but also motivational and perceptual outcomes are expected.

2.2 A New Instructional Design Model of Game-Based Learning

As is the case for all instructional design models, the above GBL model outlines the ideal instructional method since it abstracts from specific outcomes and specific groups of learners. By taking different aspects into account, a more encompassing picture of the implementation of game-based learning environments in educational settings is aimed at.

The first key element in the new GBL model is the educational game or the *game-based learning environment* (GBLE). In this chapter, both terms are used as synonyms. A GBLE contains—based on the two previous GBL models—instructional elements on the one hand (i.e., the material students have to learn from playing the game, e.g., learning content), and on the other game characteristics (i.e., the features that make the learning environment game-like, e.g., competition element). The instructional elements—in turn—consist of the learning content (i.e., the content in the learning domain the game is focusing on, e.g., proportional reasoning in mathematics) and the instructional support in the game. This support contains tools that are focusing on the learning content (e.g., correct answer feedback). The integration of these instructional elements in a GBLE is discussed more in detail in this chapter. The game characteristics contain all the features in the GBLE that make the environment game-like (e.g., storyline, interactivity, sensory stimuli). This also includes the tools that are implemented in the GBLE in order to facilitate the gameplay (e.g., tutorial with explanation about game functionalities).

The *implementation context* is the second key element of the proposed GBL model. Here, the focus is on the way the GBLE (first element) is used or applied, as GBL does not happen in a vacuum, but in a context. When two teachers use the same game in a different way, they create a different context out of which a different effect will most probably result. The implementation context is shaped by the setting (i.e., physical environment) in which the GBLE is implemented, the way the GBLE is presented to the students, the goal the educator wants to achieve, and by giving students gameplay opportunity and consequently the link or integration of the GBLE in the curriculum. Hence, a GBLE cannot be separated from the context in which it is introduced (Gee, 2011; Liu & Rojewski, 2013).

The third key element in the GBL model are the *learners* or the target group for whom the model is designed. This is done because developing learning environments requires to take into account individual differences between learners, such as differences in prior knowledge and in affective variables such as motivation (Shute & Zapata-Rivera, 2008). Furthermore, learners are active actors in learning processes (Lowyck, Elen, & Clarebout, 2004; Winne, 2004). Hence, instructional interventions (i.e., offering learners GBLEs) should not be considered as the direct cause of learner outcomes (Winne, 2004). As stated by Jonassen, Campbell, and Davidson (1994), "there is at best an indirect link between media and learning" (p. 36), and this link consists of the activities that are enabled by the media. So a direct effect of instruction on learning outcomes is not expected (Vandewaetere, Vandercruysse, & Clarebout, 2012). Rather it is the learners' perception and cognitions that affect the effectiveness of instruction. Notwithstanding the intentions from designers and teachers, the ultimate effect of instructional methods (i.e., using educational games) depends on—among others things—student' interpretation or perception of these GBLEs, rather than the GBLEs themselves. Hence, different interpretations result in different processes and products (Lowyck et al., 2004; Winne, 1987). Taken together, it is important to take the learners into account because interindividual differences in for instance perception may affect the learning results and hence the effectiveness of the intervention (Lowyck et al., 2004; Struyven, Dochy, Janssens, & Gielen, 2008).

As abovementioned, in this chapter, the first key element, i.e., the GBLE, is elaborated on, and more specific the integration of the instructional elements in a GBLE is further examined. Examining distinct elements is advocated. Up to now, no univocal definition or shared framework of educational games exists (Aldrich, 2005; Vandercruysse, Vandewaetere, & Clarebout, 2012). Based on such a framework however, educationally effective elements of a GBLE can be pinpointed, and hence, a first step towards an empirically supported conceptual research framework can be taken.

2.2.1 Game-Based Learning Environment

2.2.1.1 Instructional Elements: Learning Content

One of the instructional elements is the learning content. With respect to the learning content, two major aspects seem to be (1) the relationship between educational goals and the learning content, and (2) the suitability of integrating the learning content in a game context. Concerning the suitability of integrating the learning content in a GBLE, Malone (1980, 1981) and Malone and Lepper (1987) were the first to consider this problem of content integration in GBLEs. They state that the educational effectiveness of games depends on the way learning content is integrated into the fantasy context of the GBLE, and they propose the concepts of intrinsic and extrinsic fantasy to reveal an important distinction. This was further elaborated by Habgood,

Ainsworth, and Benford (2005) and Habgood and Ainsworth (2011), who distinguish intrinsically and extrinsically integrated games. The emphasis switched from fantasy to the core game mechanics of digital games. Following their definition, intrinsically integrated games: "(1) deliver learning material through the parts of the game that are the most fun to play, riding on the back of the flow experience produced by the game, and not interrupting or diminishing its impact and; (2) embody the learning material within the structure of the gaming world and the players' interactions with it, providing an external representation for the learning content that is explored through the core mechanics of the gameplay" (Habgood et al., 2005, p. 494).

Extrinsically integrated games separate learning components and playing components. After completing one part of the learning content, students are provided with a reward by having the chance to advance in the game without dealing with learning content (e.g., playing a subgame). Clark et al. (2011) follow this line of thought and distinguish between *conceptually integrated* and *conceptually embedded games*. In the former, learning goals are integrated into the actual gameplay mechanics, whereas this is not the case in the latter. Holbert and Wilensky propose a new design principle in addition to the conceptually integrated or embedded distinction that was made by Clark et al. (2011). They argue that games should also be representationally congruent. "Representational congruent games are construction games where the player builds and/or interacts with the game using primitives relevant to the game world to construct representations that are congruent with those used by domain experts in the real world. In such games the primitives for construction embody the content (as in conceptually integrated games), but by putting them together in personally meaningful ways, the resulting representation has meaning outside of the game." (Holbert & Wilensky, 2012, p. 371).

The integration of learning content into parts of the gameplay (i.e., intrinsic integration) ensures, in principle, game flow experiences. Because of the maintenance of flow experience, intrinsically integrated games are argued to motivate and engage players more than extrinsically integrated games (e.g., Garris et al., 2002). Clark et al. (2011) as well as Habgood and Ainsworth (2011) found that intrinsically integrated games indeed engage players (i.e., primary school children) with the learning content in the game during a longer period of time. Besides students' motivation, playing with an intrinsically integrated game might also promote learning outcomes. For instance, Habgood and Ainsworth (2011) found a higher score on a delayed mathematical post-test in the intrinsically integrated condition than in the extrinsically integrated condition. Also Echeverria, Barrios, Nussbaum, Améstica, and Leclerc (2012) found that the game in which the content was intrinsically integrated was useful for increasing students' average test results and decreasing the number of students with conceptual problems. In the study of Clark et al. (2011), the learning progress was not as extensive as hoped for, but the learning during their intrinsically integrated condition seemed to have been supported. However, in a study involving vocational secondary education (VSE) students performed by ourselves, the opposite was found.

In this study, two kinds of GBLEs were studied: an intrinsically and an extrinsically integrated mathematical GBLE. Mathematics was selected as the GBLE content since this is a well-defined domain with specific applications (i.e., not too abstract). Additionally, the math domain chosen for this study (i.e., proportional reasoning) was relevant to the curriculum of Flemish VSE. Students enrolled in this system are expected to understand the proportional reasoning language and have to be able to solve proportional reasoning problems (Vlaamse Overheid, 2010). Based on the definition of Habgood et al. (2005), the mathematical content in the intrinsically integrated game was delivered through those parts of the game that are the most fun to play and embodied within the structure of the game and the players' interactions with it. Gaming and mathematical aspects cannot be separated from each other in this version of the GBLE. This means that the gameplay is not interrupted by the mathematical learning content because it is completely interwoven with the game mechanics and storyline. In the extrinsically integrated environment, the mathematical content was not part of the core mechanics and structure of the game, but was only introduced at the beginning of every subgame as a series of separate mathematical exercises. After students have answered these items, the game continues in the same fashion as in the intrinsically integrated version of the GBLE. However, in this version of the GBLE, they do not have to make any calculations as all the mathematical items are already presented to them prior to the subgames. See Fig. 2.2 for a screenshot from the intrinsically integrated version of the GBLE and the extrinsically integrated version of the GBLE. Hence, the study focused on whether integrating mathematical content (i.e., proportional reasoning) in a particular way (intrinsic vs. extrinsic) produced different effects.

The results of the study indicated that students playing in an extrinsically integrated math game showed higher learning gain, higher motivational gain, and higher perceived usefulness than students who played with the same math game but in which the content was intrinsically integrated (Vandercruysse et al., under revision). This effect of content integration was not completely in line with the previous mentioned literature data. At the outset, it was for instance assumed that intrinsically integrating the content would stimulate students and make them outperform those students who played in the extrinsically integrated condition. A possible explanation for the surprising findings is that, integrating the learning content into the game mechanics proved to be a complex and difficult process for our particular target group, i.e., VSE students with a significant number of at-risk youths (Vandercruysse et al., under revision). Students who play with this kind of GBLE experience more difficulties in learning the content because they simultaneously have to cope with two competing demands: the educational game and the gameplay elements (Shaffer, 2004). The difficulties students experienced in the intrinsically integrated condition frustrated them to such a degree that it reduced their motivation. Additionally, the exercise formats in the extrinsically integrated GBLE showed greater similarity with the items in the pre- and post-test. This might explain why students in the intrinsically integrated condition experienced more difficulties in transferring their mathematical knowledge from one context (the game) to the next (the paper-and-pencil test) (Habgood & Ainsworth, 2011). This might suggest that the specific

Fig. 2.2 Example of a missing value problem in the intrinsic (*top*) and extrinsic (*below*) integration versions of Zeldenrust

target group playing with the GBLEs might be of influence for the effect of the content integration in GBLEs on the GBL outcomes. Furthermore, questions may arise whether or not these findings can be generalized to other target groups. In order to elucidate these concerns, future research could focus on the impact of the integration of learning content and more specifically, whether the effect is dependent on the target group under investigation.

2.2.1.2 Instructional Elements: Instructional Support

In addition to the learning content, also instructional support can be integrated as instructional element in GBLEs. Instructional support, as defined by Tobias (1982, 2009) and Tobias, Fletcher, Dai, and Wind (2011), is any type of assistance, guidance, or instruction to help students learn. Examples are scaffolds, explanations, directions, assignments, background information, monitoring tools, and planning tools (Leemkuil & de Jong, 2011; Liu & Rojewski, 2013; Tobias et al., 2011). Adding instructional support is assumed to be a necessary part of GBLEs (de Freitas & Maharg, 2014) and is expected to stimulate or facilitate students' GBL (Ke, 2009). This assumption is confirmed by previous meta-analyses conducted by Ke (2009), Lee (1999), and Wouters and van Oostendorp (2013): simulation environments and games with instructional support can improve learning.

However, Wouters and van Oostendorp (2013) emphasize that adding instructional support to games is complex since the effect is dependent on *the type of support* and *the cognitive activities* they target, among others. Moreno and Mayer (2005), for instance, investigated the role of guidance and reflection as different types of support in GBLEs. A guidance effect was found, meaning that students achieve higher transfer scores, produce fewer incorrect answers and show greater reduction of misconceptions during problem solving when guidance is added (Moreno & Mayer, 2005), while the reflection effect appeared to be less consistent. Mayer and Johnson (2010) provided evidence concerning the instructional effectiveness of reflection prompts in the form of feedback on conceptual learning (Mayer & Johnson, 2010). Another study however established that reflection only promotes retention in noninteractive environments but not in interactive environments, unless students are asked to reflect on correct program solutions rather than their own (Moreno & Mayer, 2005). Reflection prompts as support in yet another study were ascertained less promising as they did not affect students' mathematical performance and transfer (ter Vrugte, de Jong, Wouters, et al. 2015). Furthermore, Darabi, Nelson, and Seel (2009) examined the influence of supportive information (i.e., a combination of instructional strategies offered to the students in the form of text, still pictures, and graphics of critical components of a complex system to explain the nonrecurrent aspects of problem solving in the domain of chemical engineering which was the subject of the study). The results indicated a change in players' mental models after the supportive information. Supportive information in the form of conceptual clarifications seemed to be less effective in a study of Vandercruysse et al. (2016). In this study, conceptual clarifications were added to the game as instructional support. For instance, the cognitive strategies that allow students to perform the tasks in the game and hence solve the problem were offered to the students, either in an internally integrated way (i.e., the support is integrated in the GBLE for instance as an interactive tutorial; see Fig. 2.3) or an externally integrated way (i.e., the support is offered to the students apart from the GBLE on handouts). Hence, the content of the support as identical in both situations, but the way the support was integrated, differed.

Results of the study indicated that adding conceptual clarifications as instructional support in an intrinsically integrated GBLE is not recommended for VSE students. If the support is given to the students anyhow, it is advised to offer it externally because internally integrating this support leads to a decrease in performance and motivation. A possible explanation is that the support was an embedded and programmed set of information given to all the students at the start of every new

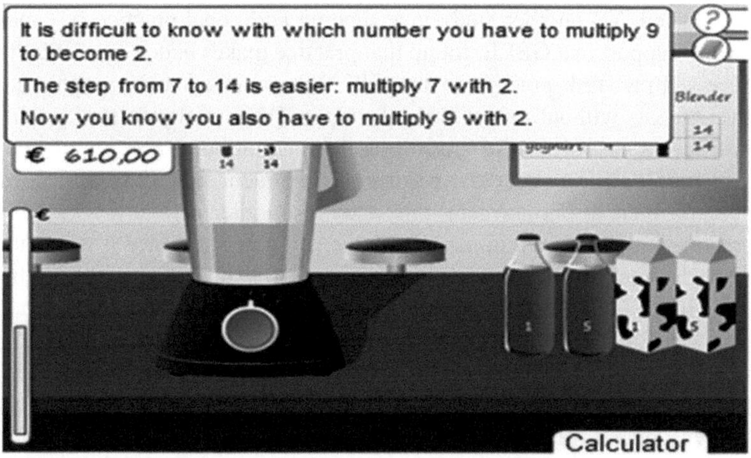

Tips for solving the problems:

Tip 1

Sometimes it is not easy to work vertically. Than it is easier to work horizontally. That is the case in the example below.

	Jug 1		Jug 2
Milk	5		11 + ?
Strawberry juice	2	x 3 →	6

Tip 2

In other problems it will be easier to work vertically instead of horizontally. Look at this example:

It is difficult to know with which number you have to multiply 2 to become 5. The step from 2 to 6 is easier: multiply 2 with 3.

Now you know you also have to multiply 5 with 3. You need 15 units of milk.

		Recipe	Milkshake
Milk	x 3 ↑	6	11 +?
Strawberry Juice		2	5

Fig. 2.3 Example of a translated part of the conceptual clarifications in the internally integrated condition and an extract from the handouts with conceptual clarifications in the externally integrated condition

subgame; irrespective of whether the players needed this information or not. Hence, the students in the ICC condition had to simultaneously cope with two forms of competing demands: the educational game with the integrated support and the gameplay elements (Shaffer, 2004). This might have been too overwhelming and have resulted in information overload. In the study of Darabi et al. (2009), problem-solving practice using a computer-based simulation was investigated as instructional support. Only a little change in mental model after problem-solving practice was established. Yet another study investigating additional practice (i.e., part-task practice) as support in a GBLE, found that practice makes better, i.e., VSE students who received part-task practice in the GBLE they played with, progressed more than the students without this additional support (Vandercruysse et al., n.d.). Part-task practice as support in this study was the integration of a series of items that provided training for a particular recurrent constituent skill (i.e., of proportional reasoning problems). Furthermore, this study also indicated that the way this part-task practice was integrated in the GBLE seemed to matter since it was found that students who received the internally integrated practice (i.e., practice that was integrated in the GBLE) improved more than the students with the same support but offered externally to the GBLE. Additional practice as support in combination with feedback was also investigated by Liu and Rojewski (2013). No effect of integrating practice and feedback in the game on participants' game enjoyment, academic achievement, or motivation was found (Liu & Rojewski, 2013). Procedural information—which was intended to aid the reflection process—had no additional value either (ter Vrugte, de Jong, Wouters, et al., 2015). Also the integration of learning units, which provide explicit instruction of the mathematical thinking strategies used in the game, did not lead to better learning outcomes between students playing with the GBLE alone and students playing with the GBLE in combination with the learning units (Broza & Barzilai, 2011). Charsky and Ressler (2011) moreover predicted that the use of concept maps would enhance the educational value of the gameplay activity; in particular students' motivation to learn through gameplay. However, the opposite happened: using conceptual scaffolds decreased students' motivation to learn through gameplay (Charsky & Ressler, 2011).

In short, the effectiveness of instructional support in games, as was also the case for the learning content, turns out to be unclear. These ambiguous findings might be a consequence of the diversity of the support (Leemkuil & de Jong, 2011; Tobias & Fletcher, 2012; Wouters & van Oostendorp, 2013). Besides the diversity of the support, another possible explanation for the ambiguity in the effects of support in GBLEs might be that two forms of integration of instructional support can be distinguished (Honey & Hilton, 2011; Ke, 2009; Liu & Rojewski, 2013). In some studies, the instructional support is internally integrated in the GBLE (e.g., Darabi et al., 2009; Johnson & Mayer, 2010; Lee, 1999; Liu & Rojewski, 2013; Mayer & Johnson, 2010; Moreno & Mayer, 2005, ter Vrugte, de Jong, Wouters, et al., 2015). In other studies, external instructional support is investigated (e.g., Barzilai & Blau, 2014; Broza & Barzilai, 2011; Charsky & Ressler, 2011). This type of support is offered to the students apart from the GBLE. There is no consensus about which type of support is the most effective. Some researchers advocate external integration (e.g., Barzilai & Blau, 2014), while others propose internal integration (e.g., Charsky

& Ressler, 2011; Liu & Rojewski, 2013). Barzilai and Blau (2014) for instance concluded from their study that external support might help learners to form connections between game knowledge and formal school knowledge, and hence improve their knowledge. Offering external support, such as concept mapping scaffolds in their study, might however also focus students' attention too much on the difficulty of the learning content and make the gameplay less self-evident (Charsky & Ressler, 2011). Therefore, Charsky and Ressler (2011) and Liu and Rojewski (2013) propose to integrate this instructional support internally into the game so it becomes an ongoing part of the gameplay. This might enhance learning and motivation. However, an important consequence of internally integrating support in games—in turn—is that, depending on the format and type of the support, it might disrupt the game flow because it demands too much processing capacity of the learner, and as in consequence the motivational nature of the game (Johnson & Mayer, 2010). Hence, depending on the type of support, the support needs to be integrated either internally or externally in order to be effective.

2.2.1.3 Game Characteristics

However, as can be derived from the instructional design model and as reviewed by Vandercruysse et al. (2012), a GBLE also contains other elements, being the game characteristics. Examples of game characteristics or elements are game rules, goals and objectives, feedback (i.e., game score), interactivity, game story, display system, and background music. Unfortunately, there is no agreement on what aspects are crucial to constitute an educational game (Vandercruysse et al., 2012). Some elements are already investigated in order to find the benefits of these elements. For instance, it was found by Richards, Fassbender, Bilgin, and Thompson (2008) that changes in pitch and tempo of the background music in educational games has no impact on learning outcomes. The display system, on the contrary, did evoke an effect on the feelings of immersion of the students (Richards et al., 2008). However, other elements evoke less conclusive findings (e.g., feedback in games; see Cornillie, 2014 for a thorough discussion about this element), or remain insufficiently investigated (e.g., game rules; Vandercruysse et al., 2012). Future research could focus more thoroughly on distinct game characteristics in order to be able to find, in addition to instructional elements, educationally effective game characteristics of a GBLE.

2.2.2 Implementation Context

Furthermore, focusing only on the design of the GBLE is too restrictive to capture the whole GBL-process. Also the implementation context is an important key element in the GBL model. The implementation context can be operationalized in different ways. For example, the implementation context might be examined by focusing on the way the game is implemented in the curriculum. Several authors (e.g., Baker &

Delacruz, 2008; Henderson, Klemes, & Eshet, 2000; Miller, Lehman, & Koedinger, 1999) already indicated the importance of integrating the game activities into the curriculum. When the gameplay is connected to the curriculum, the game is more likely to accomplish the intended instructional objectives because of the prompts that relate the game content to the curriculum. This prevents the learning processes from remaining simply inert (Tobias et al., 2011), which in turn stimulates transfer (Tobias & Fletcher, 2012). Although there appears to be consensus that games not connected to the curriculum are less likely to accomplish instructional objectives, research on the level of integration into the curriculum is largely lacking (Tobias et al., 2011). Therefore, Vandercruysse and colleagues (Vandercruysse, Desmet, et al., 2015; Vandercruysse, Van Cauwenberghe, & Elen, n.d.; Vandercruysse, van Weijnen, et al., 2015) had a try to fill the gap in this respect by exploring in three different studies possible ways of dealing with GBLEs in class and more specific, of integrating GBLEs in the curriculum.

In all three studies, curriculum was broadly interpreted as the range of activities and experiences offered at school and refers to the purposes, content, activities, and organization of the educational program enacted in the school by teachers, students, and administrators (Walker & Soltis, 1997). The first study addressed how the competition component of a game can be implemented in class by integrating rewards into the curriculum in different ways, and whether the way in which this competition is implemented matters. The focus was on competition because many researchers advocate a competition element in games (Hays, 2005). Competition is therefore incorporated in many games. However, scientific literature lacks consensus about the effectiveness of competition in games (Cheng, Wu, Liao, & Chan, 2009; Peng & Hsieh, 2012). Furthermore, the literature does not offer teachers an answer to the question about how to handle the game element competition in the classroom. Depending on the game environment, competition can be more or less emphasized, and might include rewards.

In their first study, the impact of integrating game competition in the classroom by assigning extra rewards was examined. In particular, the performance in the game led to an additional reward. In line with the findings of Hays (2005) and Tobias et al. (2011), the rewards were integrated in the curriculum in different ways. The other two studies investigated the effect of integrating a GBLE into the curriculum in yet a different way. First, a distinction was made between a strong and a weak integration by giving different instructions to the students during the intervention. The effect of instruction as support has already been investigated by Erhel and Jamet (2013). In their study, additional instruction in GBLEs seemed to have an impact on students' comprehension. More specifically, when instructions emphasized the entertainment nature of a GBLE, students performed significantly worse on memorization, in comparison to learners who received instructions focusing on the learning nature of the GBLE (Erhel & Jamet, 2013). This was further elaborated by Vandercruysse and colleagues. During the instruction, the GBLE the game content was linked in different ways to the math course in the classroom. However, this operationalization seemed insufficient and a more thorough game integration into the curriculum was carried through in the third and final study focusing on the curriculum integration. In order to attain this effective game integration, a more enhanced operationalization of the curriculum integration concept, based on the following three phases: (1) briefing, (2) playtime opportunity, and (3) debriefing (Felicia, 2011). In each of these phases, the teachers' focus is on attaining the educational goals and stimulating students' performance. Again a weak and a strong integration condition in the curriculum are investigated, using a different operationalization following the three phases of Felicia (2011). Hence, the focus is on the possible benefits for students of using a GBLE and more specifically whether integrating this GBLE in different ways (strong vs. weak) in the curriculum evokes a different effect.

The results of the three studies indicated the integration of the GBLE into students' curriculum has only minimal effect on GBL processes. None of the three studies could confirm the importance of the implementation context as a decisive variable for GBL. These findings are surprising, especially because of the importance that is often spent to the context in which learning occurs and because of the assumptions based on previous research that the curriculum integration should evoke some effect (e.g., Baker & Delacruz, 2008; Henderson et al., 2000; Miller et al., 1999).

However, the implementation context can also be operationalized alternatively. One alternative way is to investigate the setting or the classroom structure (i.e., the way the classroom activities are organized) in which the GBLE is implemented. An example is the work of Ke and Grabowski (2007). In their study, they addressed the combination of collaboration and competition in an educational math game for fifth-grade students. In particular, they explored whether computer games and collaborative learning could be used together to enrich mathematics education. The results indicated that the gaming context (collaboration or competitive) played a significant role in influencing the effect of educational gaming on affective learning outcomes. Concerning math attitudes, the collaboration condition promoted significantly more positive attitudes than the competitive or control condition. In the second study, Ke (2008) largely adopted the design of his previous study with the same target group. The cooperative condition remained the same but he divided the competitive group in two separate groups: an individualistic game playing group and a competitive game playing group. The research results indicated again that the classroom structure influenced the effects of computer games on mathematical learning outcomes and attitudes. Also ter Vrugte, de Jong, Vandercruysse, et al., 2015 explored the combination of collaboration and competition in GBL in a fully crossed design. Four conditions were examined: collaboration and competition, collaboration only, competition only, and a control group without competition and collaboration. It was found that learning effects did not differ between conditions (ter Vrugte, de Jong, Vandercruysse, et al., 2015). However, an interaction between collaboration and competition was found when students' ability levels were taken into account. Above-average students seemed to experience a positive effect of competition on domain knowledge gain in a collaborative learning situation. Below-average students showed a negative effect of competition on domain knowledge gain in a collaborative learning situation. In sum, the results of these studies indicate the importance of the classroom structure or the setting in which a GBLE is implemented, on the effectiveness of educational gaming. It shows also the importance of learner characteristics (i.e., ability level), the third key element in Fig. 2.1.

Though the different implementation choices of competition appeared of no importance in the results of the study of Vandercruysse, van Weijnen, et al. (2015), it might still be a relevant implementation context element when it is combined with collaboration. This could encourage further research based on different implementation context operationalizations. Only if the other operationalizations of the implementation context also appear to be unimportant in future research, removing the implementation context from the GBL model might be considered. Now, excluding the implementation context as important for GBL seems premature.

2.2.3 Learners

As mentioned above, individual differences between learners should be taken into account when developing learning environments because interindividual differences may affect the learning results and hence the effectiveness of the intervention. Especially, when very specific target groups are involved, this third key element might influence the GBL processes. Vandercruysse and colleagues conducted their research with VSE students. This group of students contain a significant number of at-risk students having encountered numerous unsuccessful instructional interventions and having grown resistance to traditional educational materials (ter Vrugte, de Jong, Wouters, et al., 2015). This causes among other things passivity or limited investment of effort (Placklé et al., 2014). This is aggravated by their focus on superficial instead of deep knowledge, routine instead of adaptive conceptually based approaches of learning content and by their (below) average cognitive capabilities (e.g., Cobb & McClain, 2005; Inspectie van het Onderwijs, 2009). These characteristics are specifically problematic for the acquisition of knowledge in difficult topics such as mathematics since they hinder growth in numeracy (Placklé et al., 2014). Another characteristic of VSE students is they often show a wide variety in cognitive abilities and potential (Placklé et al., 2014). It was found that students with different levels of mathematical ability are differently affected by playing with a GBLE (ter Vrugte, de Jong, Vandercruysse, et al., 2015; Vandercruysse et al., accepted). However, the findings in the studies using VSE students as target group, often deviated from assumptions based on literature and from previous empirical findings with other target groups. This might indicate the findings are target group specific: what works for VSE students does not necessarily apply to other target groups and vice versa (ter Vrugte, de Jong, Vandercruysse, et al., 2015). Because of this target group specificity, future research can focus on the specific differences between different target groups and hence investigate whether students from other education levels react similarly on the same interventions. This would enable us to pinpoint the decisive aspects of a target group for the effect of GBL and what students' variables (prior knowledge, motivation, previous gaming or school experiences, etc.) influence GBL outcomes.

2.3 Conclusion

In sum, the literature is inconclusive about the effects of instructional support and whether this support should be internally and/or externally integrated. Furthermore, the effect of intrinsically or extrinsically integrating the learning content is not yet sufficiently investigated. Nonetheless, the studies that are conducted reveal evidence for the importance of the integration of instructional elements, under certain conditions. Integration of instructional elements can happen in different ways (i.e., internal versus external or intrinsic versus extrinsic integration), and these options are decisive for the effect of the GBLE on the GBL outcomes. Further elaborating this undecidedness by examining distinct GBLE elements seems to be a promising manner to further fine-tune the effect of each of these instructional elements. Based on this kind of research, the educationally relevant elements of a GBLE can be pinpointed. This research method is advocated because up to now, no univocal definition or shared framework of educational games exists (Vandercruysse et al., 2012). Based on results of this type of research, effective parts of a GBLE can be identified and the first step towards a conceptual framework can be made. As Aldrich (2005, p. 80) stated, "Rather than thinking about games and simulations, it is more productive to think about the distinct elements." This chapter has delivered an onset towards this for the choice of instructional elements in a GBLE. More specifically, it should first be decided whether the instructional elements are offered in the GBLE (i.e., internal integration) or in addition to the GBLE (i.e., external integration). Second—after opting for internally integrating the instructional elements—it should be decided whether the instructional elements are integrated into the game mechanics (i.e., intrinsic integration) or separated from the playing components (i.e., extrinsic integration).

To conclude, it seems that the way a GBLE is designed, and more specific the interplay between the instructional elements and the game elements are an important aspect in the GBL-process. Several decisions have to be made when designing a GBLE and these decisions are of influence on GBL outcomes. In addition to the importance of the GBLE design, also the implementation context and the players (the learners) are two decisive key elements in the GBL model.

References

Aldrich, C. (2005). *Learning by doing: A comprehensive guide to simulations, computer games, and pedagogy in e-learning and other educational experiences*. San Francisco, CA: Pfeiffer.

Aleven, V., Koedinger, K., Corbett, A. T., & Perfetti, C. (2015, August). The knowledge-learning-instruction (KLI) framework: Helping to bring science into practice. In B. de Koning (Chair), *Invited SIG Symposium: Instructional design models—Do they still exist?* Symposium conducted at the 16th Biennial EARLI Conference for Research on Learning and Instruction, Limassol, Cyprus.

Baker, E. L., & Delacruz, G. C. (2008). A framework for the assessment of learning games. In H. F. O'Neil & R. S. Perez (Eds.), *Computer games and team and individual learning* (pp. 21–37). Oxford, UK: Elsevier.

Barzilai, S., & Blau, I. (2014). Scaffolding game-based learning: Impact on learning achievements, perceived learning, and game experiences. *Computers & Education, 70*, 65–79. doi:10.1016/j.compedu.2013.08.003.

Broza, O., & Barzilai, S. (2011). When the mathematics of life meets school mathematics: Playing and learning on the "my money" website. In Y. Eshet-Alkalai, A. Caspi, S. Eden, N. Geri & Y. Yair (Eds.), *Learning in the technological era: Proceedings of the Sixth Chais Conference on Instructional Technologies Research* 2011 (pp. 92–100). Ra'anana, Israel: The Open University of Israel.

Charsky, D., & Ressler, W. (2011). "Games are made for fun": Lessons on the effects of concept maps in the classroom use of computer games. *Computers & Education, 56*, 604–615. doi:10.1016/j.compedu.2010.10.001.

Cheng, H. N. H., Wu, W. M. C., Liao, C. C. Y., & Chan, T.-W. (2009). Equal opportunities tactic: Redesigning and applying competition games in classrooms. *Computers & Education, 53*, 866–876. doi:10.1016/j.compedu.2009.05.006.

Clark, D. B., Nelson, B. C., Chang, H.-Y., Martinez-Garza, M., Slack, K., & D'Angelo, C. M. (2011). Exploring Newtonian mechanics in a conceptually-integrated digital game: Comparison of learning and affective outcomes for students in Taiwan and the United States. *Computers & Education, 57*, 2178–2195. doi:10.1016/j.compedu.2011.05.007.

Cobb, P., & McClain, K. (2005). Guiding inquiry-based math learning. In R. K. Sawyer (Ed.), *The Cambridge handbook of the learning sciences* (pp. 171–186). Cambridge, UK: Cambridge University Press. doi:10.1017/CBO9780511816833.012.

Cornillie, F. (2014). *Adventures in red ink. Effectiveness of corrective feedback in digital game-based language learning.* Unpublished doctoral dissertation, Katholieke Universiteit Leuven, Belgium.

Darabi, A. A., Nelson, D. W., & Seel, N. M. (2009). Progression of mental models throughout the phases of a computer-based instructional simulation: Supportive information, practice, and performance. *Computers in Human Behavior, 25*, 723–730. doi:10.1016/j.chb.2009.01.009.

de Freitas, S., & Maharg, P. (2014). Series editors introduction. In N. Whitton (Ed.), *Digital games and learning: Research and theory* (pp. xiii–xiv). New York: Routledge.

Echeverria, A., Barrios, E., Nussbaum, M., Améstica, M., & Leclerc, S. (2012). The atomic intrinsic integration approach: A structured methodology for the design of games for the conceptual understanding of physics. *Computers & Education, 59*, 806–816. doi:10.1016/j.compedu.2012.03.025.

Erhel, S., & Jamet, E. (2013). Digital game-based learning: Impact of instructions and feedback on motivation and learning effectiveness. *Computers & Education, 67*, 156–167. doi:10.1016/j.compedu.2013.02.019.

Felicia, P. (2011). *How can digital games be used to teach the school curriculum.* Retrieved from http://linked.eun.org/c/document_library/get_file?p_l_id=22779&folderId=24664&name=DLFE-783.pdf

Gagné, R. M., Briggs, L. J., & Wager, W. W. (1992). *Principles of instructional design* (4th ed.). Forth Worth, TX: Harcourt Brace Jovanovich College Publishers.

Garris, R., Ahlers, R., & Driskell, J. E. (2002). Games, motivation, and learning: A research and practice model. *Simulation & Gaming, 33*, 441–467. doi:10.1177/1046878102238607.

Gee, J. P. (2011). Reflections on empirical evidence on games and learning. In S. Tobias & J. D. Fletcher (Eds.), *Computer games and instruction* (pp. 223–232). Charlotte, NC: Information Age Publishing.

Habgood, M. P. J., & Ainsworth, S. E. (2011). Motivating children to learn effectively: Exploring the value of intrinsic integration in educational games. *Journal of the Learning Sciences, 20*, 169–206. doi:10.1080/10508406.2010.508029.

Habgood, M. P. J., Ainsworth, S. E., & Benford, S. (2005). Endogenous fantasy and learning in digital games. *Simulation & Gaming, 36*, 483–498. doi:10.1177/1046878105282276.

Hays, R. T. (2005). *The effectiveness of instructional games: A literature review and discussion* (Technical Report No. 2005-004). Orlando, FL: Naval Air Warfare Center Training Systems Division.

Henderson, L., Klemes, J., & Eshet, Y. (2000). Just playing a game? Educational simulation software and cognitive outcomes. *Journal of Educational Computing Research, 22*, 105–129. doi:10.2190/EPJT-AHYQ-1LAJ-U8WK.

Holbert, N., & Wilensky, U. (2012). *Representational congruence: Connecting video game experiences to the design and use of formal representations*. Proceedings of Constructionism 2012.

Honey, M. A., & Hilton, M. (Eds.). (2011). *Learning science through computer games and simulations*. Washington, DC: The National Academies Press.

Inspectie van het Onderwijs. (2009). *De staat van het onderwijs. Onderwijsverslag 2007/2008* [*The state of education. Education report (2007/2008)*]. De Meern: Inspectie van het Onderwijs.

Johnson, C. I., & Mayer, R. E. (2010). Applying the self-explanation principle to multimedia learning in a computer-based game-like environment. *Computers in Human Behavior, 26*, 1246–1252. doi:10.1016/j.chb.2010.03.025.

Jonassen, D. H., Campbell, J. P., & Davidson, M. E. (1994). Learning with media: Restructuring the debate. *Educational Technology Research & Development, 42*, 31–39. doi:10.1007/BF02299089.

Ke, F. (2008). Computer games application within alternative classroom goal structures: Cognitive, metacognitive, and affective evaluation. *Educational Technology Research and Development, 56*, 539–556. doi:10.1007/s11423-008-9086-5.

Ke, F. (2009). A qualitative meta-analysis of computer games as learning tools. In R. E. Ferdig (Ed.), *Handbook of research on effective electronic gaming in education* (pp. 1–32). Hershey, PA: IGI Global. doi:10.4018/978-1-59904-808-6.ch001.

Ke, F., & Grabowski, B. (2007). Gameplaying for maths learning: Cooperative or not? *British Journal of Educational Technology, 38*, 249–259. doi:10.1111/j.1467-8535.2006.00593.x.

Lee, J. (1999). Effectiveness of computer-based instructional simulation: A meta analysis. *International Journal of Instructional Media, 26*(1), 71–85.

Leemkuil, H., & de Jong, T. (2011). Instructional support in games. In S. Tobias & J. D. Fletcher (Eds.), *Computer games and instruction* (pp. 353–369). Charlotte, NC: Information Age Publishing Inc.

Liu, Y., & Rojewski, J. W. (2013). Effects of instructional support in game-based learning: An analysis of educational games from design and application perspectives. In R. McBride & M. Searson (Eds.), *Proceedings of Society for Information Technology & Teacher Education International Conference 2013* (pp. 43–50). Chesapeake, VA: Association for the Advancement of Computing in Education (AACE).

Lowyck, J., Elen, J., & Clarebout, G. (2004). Instructional conceptions: Analyses from an instructional design perspective. *International Journal of Educational Research, 41*, 429–444. doi:10.1016/j.ijer.2005.08.010.

Malone, T. W. (1980). What makes things fun to learn? Heuristics for designing instructional computer games. In *Proceedings of the 3rd ACM SIGSMALL Symposium and the 1st SIGPC Symposium* (pp. 162–169). doi: 10.1145/800088.802839.

Malone, T. W. (1981). Toward a theory of intrinsically motivating instruction. *Cognitive Science, 5*, 333–369. doi:10.1016/S0364-0213(81)80017-1.

Malone, T. W., & Lepper, M. R. (1987). Making learning fun: A taxonomy of intrinsic motivation for learning. In R. E. Snow & M. J. Farr (Eds.), *Aptitude, learning and instruction. Vol. 3: Conative and affective process analysis* (pp. 223–253). Hillsdale, NJ: Lawrence Erlbaum Associates Publishers.

Mayer, R. E., & Johnson, C. I. (2010). Adding instructional features that promote learning in a game-like environment. *Journal of Educational Computing Research, 42*, 241–265. doi:10.2190/EC.42.3.a.

Merrill, M. D. (2002). First principles of instruction. *Educational Technology Research and Development, 50*(3), 43–59. doi:10.1007/BF02505024.

Miller, C. S., Lehman, J. F., & Koedinger, K. R. (1999). Goals and learning in micro worlds. *Cognitive Science, 23*, 305–336. doi:10.1207/s15516709cog2303_2.

Moreno, R., & Mayer, R. E. (2005). Role of guidance, reflection, and interactivity in an agent-based multimedia game. *Journal of Educational Psychology, 97*, 117–128. doi:10.1037/0022-0663.97.1.117.

Peng, W., & Hsieh, G. (2012). The influence of competition, cooperation, and player relationship in a motor performance centered computer game. *Computers in Human Behavior, 28*, 2100–2106. doi:10.1016/j.chb.2012.06.014.

Placklé, I., Könings, K. D., Jacquet, W., Struyven, K., Libotton, A., van Merriënboer, J. J. G., et al.. (2014). Students' preferred characteristics of learning environments in vocational secondary education. *International Journal for Research in Vocational Education and Training (IJRVET), 1*, 107–124. doi: 10.13152/IJRVET.1.2.2.

Reigeluth, C. M., Merrill, M. D., Wilson, B. G., & Spiller, R. T. (1980). The elaboration theory of instruction: A model for sequencing and synthesizing instruction. *Instructional Science, 9*, 195–219. doi:10.1007/BF00177327.

Richards, D., Fassbender, E., Bilgin, A., & Thompson, W. F. (2008). An investigation of the role of background music in IVW's for learning. *ALT-J: Research in Learning Technology, 16*, 231–244.

Ryder, M. (2015). *Instructional design models and methods.* Retrieved from http://www.instructionaldesigncentral.com/htm/IDC_instructionaldesignmodels.htm#gagne

Shaffer, D. W. (2004). Pedagogical praxis: The professions as models for postindustrial education. *Teachers College Record, 106*, 1401–1421.

Shute, V. J., & Zapata-Rivera, D. (2008). Adaptive technologies. In J. M. Spector, M. D. Merrill, J. J. G. van Merriënboer, & M. P. Driscoll (Eds.), *Handbook of research on educational communication and technology* (3rd ed., pp. 277–294). New York, NY: Taylor and Francis.

Struyven, K., Dochy, F., Janssens, S., & Gielen, S. (2008). Students' experiences with contrasting learning environments: The added value of students' perceptions. *Learning Environments Research, 11*, 83–109. doi:10.1007/s10984-008-9041-8.

ter Vrugte, J., de Jong, T., Vandercruysse, S., Wouters, P., van Oostendorp, H., & Elen, J. (2015). How competition and heterogeneous collaboration interact in prevocational game-based mathematics education. *Computers & Education, 89*, 42–52. doi:10.1016/j.compedu.2015.08.010.

ter Vrugte, J., de Jong, T., Wouters, P., Vandercruysse, S., Elen, J., & van Oostendorp, H. (2015). When a game supports prevocational math education but integrated reflection does not. *Journal of Computer Assisted Learning, 31*, 462–480. doi:10.1111/jcal.12104.

Tobias, S. (1982). When do instructional methods make a difference? *Educational Researcher, 11*(4), 4–9.

Tobias, S. (2009). An eclectic appraisal of the success or failure of constructivist instruction. In S. Tobias & T. D. Duffy (Eds.), *Constructivist theory applied to education: Success or failure?* (pp. 335–350). New York: Routledge, Taylor and Francis.

Tobias, S., & Fletcher, J. D. (2012). Learning from computer games: A research review. In S. De Wannemacker, S. Vandercruysse, & G. Clarebout, (Eds.), *Serious games: The challenge.* (Vol. CCIS 280, pp. 6–18). Berlin, Germany: Springer. doi: 10.1007/978-3-642-33814-4_2.

Tobias, S., Fletcher, J. D., Dai, D. Y., & Wind, A. P. (2011). Review of research on computer games. In S. Tobias & J. D. Fletcher (Eds.), *Computer games and instruction* (pp. 127–221). Charlotte, NC: Information Age Publishing Inc.

Vandercruysse, S., ter Vrugte, J., de Jong, T., Wouters, P., van Oostendorp, H., Verschaffel, L., et al. (accepted). Content integration as a factor in math game effectiveness. *Educational Technology Research & Development.*

Vandercruysse, S., Desmet, E., Vandewaetere, M., & Elen, J. (2015). Integration in the curriculum as a factor in math-game effectiveness. In J. Torbeyns, E. Lehtinen, & J. Elen (Eds.), *Describing and studying domain-specific serious games* (pp. 133–153). Cham, Switzerland: Springer International Publishing AG. doi:10.1007/978-3-319-20276-1_9.

Vandercruysse, S., ter Vrugte, J., de Jong, T., Wouters, P., van Oostendorp, H., Verschaffel, L., et al. (2016). The effectiveness of a math game: The impact of integrating conceptual clarification as support. *Computer in Human Behaviour.* 64, 21–33.

Vandercruysse, S., ter Vrugte, J., de Jong, T., Wouters, P., van Oostendorp, H., Verschaffel, L., et al. (n.d.). The effectiveness of a math game: The impact of integrating part task practice as support. *Computers & Education.*

Vandercruysse, S., Van Cauwenberghe, V., & Elen, J. (n.d.). The effectiveness of game-based learning: The impact of curriculum integration. *Journal of Curriculum Studies.*

Vandercruysse, S., van Weijnen, S., Vandewaetere, M., & Elen, J. (2015). Competitie als game element integreren in de BSO-klaspraktijk. [Integrating competition as game element in the vocational secondary classroom]. *Pedagogische Studiën, 92*, 179–201.

Vandercruysse, S., Vandewaetere, M., & Clarebout, G. (2012). Game-based learning: A review on the effectiveness of educational games. In M. Cruz-Cunha (Ed.), *Handbook of research on serious games as educational, business, and research tools* (pp. 628–647). Hershey, PA: IGI Global. doi:10.4018/978-1-4666-0149-9.ch032.

Vandewaetere, M., Vandercruysse, S., & Clarebout, G. (2012). Learners' perceptions and illusions of adaptivity in compute-based learning environments. *Educational Technology Research and Development, 60*, 307–324. doi:10.1007/s11423-011-9225-2.

Vlaamse Overheid. (2010). *Project Algemene Vakken. Concretisering eindtermen. Secundair onderwijs—Tweede graad BSO* [*Project General Subjects. Reifying the attainment targets. Secondary education—Second grade VSE*]. Brussel: Vlaams Ministerie van Onderwijs en Vorming.

Walker, D. F., & Soltis, J. F. (1997). *Curriculum and aims*. New York, NY: Teachers College Press.

Winne, P. H. (1987). Why process-product research cannot explain process-product finding and a proposed remedy: the cognitive mediational paradigm. *Teaching and Teacher Education, 3*, 333–356. doi:10.1016/0742-051X(87)90025-4.

Winne, P. H. (2004). Students' calibration of knowledge and learning processes: Implications for designing powerful software learning environments. *International Journal of Educational Research, 41*, 466–488. doi:10.1016/j.ijer.2005.08.012.

Wouters, P., & van Oostendorp, H. (2013). A meta-analytic review of the role of instructional support in game-based learning. *Computers & Education, 60*, 412–425. doi:10.1016/j.compedu.2012.07.018.

Chapter 3
Two Systems, Two Stances: A Novel Theoretical Framework for Model-Based Learning in Digital Games

Mario M. Martinez-Garza and Douglas B. Clark

Abstract Recent reviews of quantitative research suggest that some but not all digital games add value when used as pedagogical tools. A more sophisticated cognitive theory of learning is required to guide the advance of educational games through improvements in design, scaffolding, and assessments. This chapter extends and improves existing mental model-based hypotheses about learning in games, particularly in terms of science learning and seeks to conceptualize simulation and game-based learning within a more general two-system theory of human cognition.

Keywords Educational games • Theory of learning • Mental models • Two-system theory of cognition

3.1 Introduction

Players of digital games present a remarkable duality. On the one hand, players of games often seem to sit in absorption, interacting with a complex digital game in an automatic and nearly effortless way. Observing an individual person at play, it might appear at times that the person is doing little more than reacting to stimuli, rarely demonstrating anything that resembles thinking or learning. On the other hand, players are also deeply reflective about the games they play. This is most visible in the online spaces where communities of players coalesce. In these spaces, we find that games are objects of analysis, inquiry, commentary, interpretation, and reinterpretation. As a means of participation, these activities are in some ways as valid as playing the games themselves.

M.M. Martinez-Garza (✉)• D.B. Clark
Department of Teaching and Learning, Peabody College VanderBilt University, 230 Appleton Place, Nashville, TN 37203-5721, USA
e-mail: mario.martinez@berkeley.edu

P. Wouters, H. van Oostendorp (eds.), *Instructional Techniques to Facilitate Learning and Motivation of Serious Games*, Advances in Game-Based Learning, DOI 10.1007/978-3-319-39298-1_3

This duality becomes more salient, and problematic, in the case of educational games. When the stated goal of a game is that its players learn particular content or concepts, the prevalence of automatic and reactive forms of play make it difficult to argue that anything is actually being learned beyond the performance requirements of the game itself. Ideally, the goal is that students engage with educational games in a thoughtful and purposeful way, using these games as tools for organizing their insights and furthering their understanding of the concepts the game is intended to teach. This is not to say that no learning can happen through automatic forms of play; it is certainly true that digital games can support learning of, for example, facts and procedural skills. A more challenging goal is to support learning of the causal relationships and functional properties that feature prominently in science education. These concepts may exist in learners' minds in an intuitive form, and might thus be reinforced through intuitive play. However, the process of normalizing and organizing these intuitions into more formalized modes likely requires the learner to engage with digital games in a more deliberate, analytical manner.

How can we make sense of this duality? How can we reconcile the two modes of digital gaming, the automatic and the thoughtful, into a coherent framework that explains how and what people learn from these games? How can we promote forms of play and learning that are more closely aligned with the goals of education, particularly science education?

This chapter explains what we call the Two Stance Framework of game-based learning, which is our attempt to answer these questions from a cognitive perspective.[1] The Two Stance Framework of game-based learning is an instantiation of a general theory of human cognition, *the two-system theory*, in the domain of digital games. The Two Stance Framework, or 2SM for short, envisions that players of digital games shuttle between two distinct epistemic stances: (1) a "learning" stance, which is directed toward making sense of the games' rules, the entities and relationships it portrays and (2) a "playing" stance that is geared toward optimizing in-game performance and continuing play. Furthermore, we conjecture that people develop two distinct forms of knowledge through interacting with a game: (1) an understanding of the network of entities and causal structure of the interactive model and (2) a store of practical knowledge of how to act effectively within the game. The overall goals of this chapter are to present the 2SM framework, review the theoretical bases it builds upon, and highlight the salient questions of game-based learning and implications for design that it helps to clarify.

[1] Cognitive perspectives are by no means the only productive ways to examine game-based learning. Sociocultural, situative, and embodied perspectives are also relevant and have extensive research histories. Our choice to use a cognitive lens is based on the persuasiveness of the available scholarship in light of the overarching goals of the 2SM.

3.2 Scope of Inquiry

We limit our analysis to digital games designed around *interactive models* where game play focuses on manipulating the interactive model and improving performance in the game is dependent on developing a more sophisticated understanding of the interactive model. Not all digital games (educational or leisure) can be fairly said to include such a model. In some cases, games rely on more simple ludic forms (e.g., those that challenge a player's dexterity, memory, reflexes) to provide some logic to the events and entities portrayed in the game. On the other hand, some games require the player to manipulate interactive models that are more sophisticated than those featured in simulations typically used in formal learning environments. These leisure games can focus on formal science concepts, for example, where the central challenge involves navigation or manipulation based on Newtonian relationships (e.g., *Kerbal Space Program*) or the exploration, identification, and exploitation of mineral and ecological resources (e.g., *Dwarf Fortress*). If we also consider interactive models that represent complex networks of entities and relationships which do not necessarily correspond to real-world phenomena, then the domain of digital games that forefront interactive models becomes even larger. That domain now includes not only science-based examples, but also games like *Civilization* or *SimCity* in which the player develops cities and empires in the context of elaborate networks of resource production and consumption that mimic those in the real world, or *EVE Online*, where players collectively participate in a laissez-faire economy directly caused by players' interactions within the game. In all of these case, the interactive model forms a "learnable core" that a player comes to master and understand through play.

3.2.1 The Two-System Theory of Cognition

The Two-Stance Framework seeks to explain the fact that players of digital games seem to be equally capable of quick decisions made with minimal information as well as slow deliberations that include a lot of data. This duality is not, however, unique to digital games: it is a feature of human cognition in general. Digital gaming is simply one realm in which this feature has notable effects. Thus, general theories of human cognition that shed light on the aforementioned duality are especially relevant. Both elements of this duality in reasoning and choice have their own lines of research, yet the framework that harmonizes them is of comparatively recent mint.

Contemporary scholarship has produced two broad perspectives for explaining human reasoning: classical rationality and bounded rationality. Classical rationality builds on Aristotelian and Hegelian notions of how the mind operates (see Chater & Oaksford, 1998; Smith, Langston, & Nisbett, 1992; Stenning & Van Lambalgen, 2008); it is also called *unbounded rationality*, as it describes logic-based processes of reasoning with little or no regard for how time-consuming or information intensive these processes might be. The argument implicit in classical rationality is that, to the degree that people are capable of making optimal choices, the mind must be capable of whatever computations or information processes are required. In contrast, the bounded rationality approach suggests that it may be more appropriate to acknowledge the peculiar properties and limits of the mind and environment; thus, optimal performance is not a requirement or assumption of bounded rationality approaches (Gigerenzer & Goldstein, 1996; Gigerenzer & Selten, 2001). Bounded rationality therefore not only assumes that people are limited in computational power, time, and knowledge, but also that each environment varies in affordances for making information available. Simon (1956) proposed what is arguably the predominant theory of bounded rationality, which he called "satisficing." In this approach, people's minds reason in a way that makes mostly correct choices, but within the constraints of their limited abilities to search for and process information. These satisficing mechanisms might have biological origins that are evolutionarily ancient (e.g., Stanovich, 1999). Although critics claim that unbounded rationality seems unworkable and satisficing seems reductionist, it is clear there is a place in theories of cognition for quick processes that work with limited information (bounded rationality) as well as more resource-intensive, rule-following, information-rich processes (unbounded rationality).

Thus a synthetic approach, called the dual-process theory of cognition, aims to explain the simultaneous existence of, and the interplay between, the process of mind best described as "intuition," or effortless thought, and the more deliberate, purposeful activity usually called "reasoning" (Chaiken, 1980; Epstein, 1994; Nisbett, Peng, Choi, & Norenzayan, 2001; Sloman, 1996; Stanovich, 1999). These modes of cognition are neutrally labeled in the literature as System 1 and System 2, respectively. The former is described as fast, automatic, associative, emotional, and opaque; the latter as slower, controlled, serial, and self-aware (see Evans, 2008; Kahneman, 2003).

There are two main strands of the dual-processing theory; they differ in the hypothesized relationship between Systems 1 and 2. In one strand (e.g., Sloman, 1996), both forms of processing are active in parallel, and in the other strand (e.g., Evans, 2006), they act sequentially and selectively depending on context. Research in both strands agrees that System 1 processing is more common in everyday tasks than System 2. The difference in effort required by these two systems indicates that the processes involved in System 1 are similar to basic performance-oriented computations that the mind has evolved to make. The biological origins of System 1, which are postulated to be shared with other animals, is a recurring theme in two-system theories of cognition (Evans, 2008). In parallel-processing theories of cognition, the preference for System 1 processes is explained as a strategy to minimize cognitive effort, i.e., the "cognitive miser" of Fiske and Taylor (1991). In sequential-processing theories, System 1 is seen as the default mode of cognition, with System 2 acting in a more supervisory, inhibitory, and/or interventionist role (see Stanovich, 1999).

In summary, the two-system theory of cognition seeks to explain the fact that people seem to be equally capable of both quick decisions made with minimal information and also slow deliberations that include a lot of data. The evidence supporting the two-system theory of cognition is extensive and persuasive (e.g., Evans, 2003; Sloman, 1996). While the two-system theory of cognition is still undergoing extensive refinement and conceptual clarification (Evans & Stanovich, 2013), it provides a useful structure for understanding how students reason with interactive models and what they may learn from them. The 2SM expands the two-system theory with two additional features: the Player/Learner stance dichotomy and the contraposition between interactive models and second-order models. These features are explained further in the following section.

3.3 Proposed Structure and Function of the 2SM

3.3.1 Two Models, Two Stances

Before discussing the Two Stance Framework, we distinguish between the three types of models we will encounter while exploring the 2SM. These models vary substantially in terms of their domain, function, and where they are said to reside. We refer to most of these entities as "models" for simplicity and to accurately convey the terminology used by other researchers. However, we will define and differentiate these terms as much as possible to avoid confusion and enhance clarity. The resulting constructs are summarized and compared in Table 3.1.

3.3.2 Models in the 2SM

Models have a well-established history in both science and science education, and the term encompasses several different ontologies. The two most germane to this inquiry are what we call the "external model" and the "interactive model." By "external model," we mean the formal abstraction of the scientific phenomena of interest. These abstractions have both explicatory and predictive power and as such are frequent targets of science instruction (Clement, 2000; see also Lehrer & Schauble, 2005). By "interactive model," we refer to programmed software instantiations of these formal abstractions. Both external models and interactive models have

Table 3.1 Taxonomy of model constructs

	What is represented	Goal	Applicability	Uses
External model	Physical reality	Explanation and prediction	External phenomena	Describing phenomena and predicting possible future states
Interactive model	Physical reality	Representational fidelity	Virtual phenomena	Pedagogical demonstration Supporting exploration
Mental model	Physical reality	Understanding	External phenomena	Problem-solving
Second-order model	Hybrid	Risk-free experimentation	Virtual and external phenomena	Testing and revising game strategies

pedagogical value, and the ability to present them in a coherent yet engaging way (as in, for example, PhET interactive simulations, Perkins et al., 2006) is an important affordance for science education.

A person who interacts with a game constructs some mental analogue as a necessary step in understanding its inner workings. The existence of this analogue is widely recognized by psychology and educational research and is most often called a mental model (see Mayer, 2005). An accurate, flexible mental model is hypothesized to form a crucial aspect of science expertise (Chi, Glaser, & Rees, 1981), and thus, refining such a model is a focal pursuit of both science education and games for science learning research. The exact nature of this mental model, however, is somewhat underspecified (Doyle & Ford, 1998). The mainline view is represented by Vosniadou and Brewer's (1994) definition of a mental model as "a mental representation [whose] structure is analogue to the states of the world that it represents" (p. 125).

Mental models are frequently featured in cognitivist explanations of the causal mechanisms behind science learning in digital games. In the literature on the use of games to support the goals of science education reviewed by Clark, Nelson, Sengupta, and D'Angelo (2009), 21 of the 83 papers cited contained some causal explanation for how games may help students learn science, and of those, 11 explicitly mentioned mental models. In these papers, both play and learning are envisioned to be driven by a unitary *mental model* that a learner forms in order to handle the challenges presented by the game. This mental model grows in sophistication and completeness through play and remains available to the student as a tool for problem-solving. The explanations that researchers propose for digital learning vary somewhat, but they converge on a form best articulated by Moreno and Mayer (2000): "When students try hard to make sense of the presented material, they form a coherent mental model that enables them to apply what they learned to challenging new problem-solving situations (Mayer & Wittrock, 1996)" (p. 727).

In other words, the proposed mechanism for how students learn science from games is a two-step process. In the first step, students purposefully investigate the digital environment, and "try hard to make sense" of the entities, relationships, and regularities of the portrayed reality. The product of their effort is a "coherent mental model." In the second step, students evaluate situations and solve problems in some future moment using this very same mental model. Notably, the proposition that presenting students with an external model in the hopes that through engagement they will develop a parallel internal mental model also underlies much research on science learning in labs, inquiries, simulations, and other activities.

This explanation implies that the mental models that students create are relatively persistent, flexible, and context independent. In other words, students form mental models that are fixed in long-term memory, that can be applied to help solve problems of various forms, and that can be transferred beyond the context in which they are formed. It is fundamentally true that people use mental models for all kinds of cognitive processes, including inference, judgment, and prediction (Johnson-Laird, 1983), and digital games should be no exception. It is also likely true that these mental models originate in inferences made from repeated experience (Gigerenzer, Hoffrage, & Kleinbölting, 1991; see also Hasher & Zacks, 1979).

Some researchers propose additional properties of the mental models that students develop in the course of interacting with a digital game. For example, Rosenbaum, Klopfer, and Perry (2007) equate increased understanding of a system with a more sophisticated mental model. Marino, Basham, and Beecher (2011) claim that video games promote mental models that have "coherence," using the term in the sense used by McNamara and Shapiro (2005), i.e., that mental models are well-structured representations built from a combination of the person's prior knowledge and the relevant conceptual elements from the game. The audiovisual affordances of the learning environment that aid the formation of mental models are also noted by Clark and Jorde (2004), Taylor, Pountney, and Baskett (2008), Jones, Minogue, Tretter, Negishi, and Taylor (2006), and Moreno and Mayer (2005). The mental model that a student forms is also envisioned by various authors as a tool for understanding and testing scientific theories (e.g., Anderson & Barnett, 2011; Bekebrede & Mayer, 2006; Li, 2010).

The underlying assumption is that as the learner's mental model develops, it grows to resemble the external model, so both models eventually encompass a similar set of concepts and relationships. If so, then in the case of digital games the development of the mental model is mediated by the interactive model, and grounded in the objects of the game's simulated world. A plausible conjecture is that the developing mental model also accounts for the interactive model as a result of playing the game. So, for example, a player of *Angry Birds* forms a mental model that encompasses both game-specific objects (i.e., birds flung from a slingshot) in parabolic trajectories and also real-world entities that exhibit similar properties, e.g., thrown balls. If interactive models found in digital games exhibit this sort of "transitive property," that is, a capacity to stand both for real-world objects and game-specific representations, and if the resulting mental models have both explicatory and predictive capacity (as external models do), then it is clear why they would be valuable as pedagogical tools.

However, if developing mental models only approximate interactive models, and not external models (or to the objects of these models, namely external reality), then things grow somewhat more muddled. So in the previous example, it may be the case that a person playing *Angry Birds* builds a mental analogue of how digital birds are launched, how far they fly, and the shape of their trajectories, and yet this mental analogue might have no consistent reference to any object in the real world. Some research on games for learning assumes that playing games informs mental models that map seamlessly onto external reality, yet this proposition seems problematic. First, bridging between in-game and real-world entities requires players to engage in a process of abstraction (or "high-road transfer") that is generally difficult and not often observed (see Detterman, 1993). Second, not all interactive models are wholly accurate representations of the "real world." Interactive models may be sophisticated, yet need only be "sufficiently representative of the system to yield the desired information" (Apostel, 1961, p. 126) to function as the core of a game. So players of games would also have to negotiate the occasional mismatch between in-game representation and the information from their own senses (e.g., an interactive physics model might not include friction whereas the real world does).

A more tenable proposition might be that players of digital games develop mental models that encompass the game's interactive model but only partially extend beyond the game to describe external reality. Thus, the resulting mental models is a parsimonious response to the task demands of playing the game itself. The systems control literature, a domain perhaps more akin to gaming than science education, offers some guidance as to why this might be the case. In this literature, the purpose of mental models is to generate descriptions of the system's purpose, form explanations of the system's functioning and the observed system states, and make predictions of future system states (Rouse & Morris, 1986). Any mental model that a person forms in response to a digital game has at least some control orientation; in other words, the mental model must provide some faculty that allows a player to exercise control over the game. Crucially, a person playing a digital game does not necessarily craft a mental model that accounts for the entire capabilities of the game's interactive model because such a comprehensive model is not required. All a mental model needs to provide the user is a sense that he or she understands the game and is in control of it. Yet the mental model cannot *only* have a control orientation because not all of a game's phenomena flow from the player's control. Aside from game events that are guided by the underlying interactive model, games also comprise a large variety of elements that are intended to structure the experience. These include, for instance, elements regarding the game's interface, its characters and landscapes, narrative events, and interactions between game entities that exhibit regularity but are not subject to a player's input. These elements must necessarily form part of the person's mental model, even when they do not contribute to a control orientation.

Thus, we argue that the mental models that arise in a player's thinking when playing digital games are not exactly the same as the mental models that are hypothesized to influence science learning. Rather, the mental analogues that influence game play have a great deal of hybridity. They are in some measure informed both by external reality and by the representations of the game's fictional world as driven by its interactive model, even when those two stand in conflict. They are also more control oriented and more limited in scope, and thus may be less comprehensive, less accurate, and less consistent than the mental models as described by some researchers. The mental analogues are also strongly influenced by a person's previous game-playing experiences. To represent this departure from a mainline view of mental models, and to express this hybrid structure, we henceforth refer to the mental analogue as a *second-order model* rather than mental model. Table 3.1 (below) clarifies the distinctions between the model-related constructs from the perspective of the 2SM.

3.3.3 Stances in the 2SM

Another necessary clarification for the 2SM involves stances. For the purposes of the 2SM, the term "stance" refers to the state-of-mind or stance a person takes with regard to his or her second-order model, or the game itself. When creating, refining, or applying a second-order model, a person might have two distinct goals. The first is to understand the formal structure of the interactive model and the affordances of the causal relationships it represents. The second is to use the second-order model as a laboratory where actions can be planned and evaluated in terms of their effectiveness at creating a desired state. These two sets of goals imply qualitatively different forms of thinking. A user in the learning stance might purposefully investigate the game in search of information that confirms or disconfirms his or her understanding. A user in the control stance might engage in developing and adopting control strategies, selecting proper actions, and querying the game to determine whether or not these actions lead to desired results (for further description of the "control stance," see Veldhuyzen & Stassen, 1977). To distinguish between these two stances, we envision the person seeking to further understand the interactive model as engaging in a "learner stance," and the person in the control stance who is actively working toward a goal as engaging in a "player stance."

These stances are epistemological in function; they offer collections of resources that a person can use to decide how best to think about the game as they experience it. This definition follows from Hammer and Elby (2003), who conceptualize "naïve epistemologies" as collections of resources, each activated in appropriate and familiar contexts, e.g., "motion requires force" and "maintaining agency" as resources useful for think about flying objects (see also Elby & Hammer, 2001; Hammer, Elby, Scherr, & Redish, 2005). Hammer and Elby do not provide an exhaustive taxonomy of resources, but what these resources all have in common is their relation to context. According to Hammer and Elby, resources are activated in a way that is *context sensitive*; in the process of thinking epistemologically, students select from the resources they have the ones that appear to be appropriate and productive in the current context. If the stances of the 2SM have this same texture, then people who play games assemble their stances based on (a) their knowledge resources, as provided by their personal experiences playing games, and (b) the cues that current context offers to activate those resources. The 2SM only describes the form of the stances generally; it is more concerned with describing the stances' functioning and how they interact to influence play and learning. The broad characteristics of the stances are given in Table 3.2 (below).

In the 2SM framework, the Player stance contains, as one of many component resources, the person's motivation to engage. Ryan, Rigby, and Przybylski (2006) found that motivation to engage with video games is strongly predicated on subjec-

Table 3.2 Characteristics of the player and learner stances

	Processes	Goals	Disrupted by
Player stance	Application of execution rules, evaluation of rule effectiveness after the fact	Achieve desired psychological states, maintain agency	Boredom, frustration
Learner stance	Definition and refinement of strategic rules, testing their effectiveness	Signal understanding of the interactive model, bolster agency, and self-efficacy	Interactive model that is inscrutable, inconsistent, abstruse

tive experiences of autonomy, competence, and relatedness. These three constructs are described as basic human needs in self-determination theory (Deci & Ryan, 1985). The basic needs explanation accounts for why people choose to play *a particular game* as well as why they choose to play *any* games at all. Several researchers of games, for example, have investigated why individuals choose to play one game over another (e.g., Bartle, 1996; Sherry, Lucas, Greenberg, & Lachlan, 2006; Yee, 2006), and the general finding is that people choose their games based on an individual preference for certain psychological states (e.g., challenge, competition, social interaction, or a combination thereof). Regardless of whether he or she is seeking to get a good grade, win, perfect a moment of performance, connect with fellow players, or simply learn about the underlying formal abstraction of the scientific phenomenon, the player is driven by the expectation and realization of experiences that trigger subjective feelings of autonomy, competence, and relatedness.

Insofar as the person is able to find these experiences, then he or she will seek to continue to engage with the game. Yet in order to continue to engage (and access more of these experiences), the player must at times probe the complexities of the game's underlying interactive model, learn them, and learn from them. Learning is thus not merely a residual by-product of engagement, but a necessary activity for free, effective, and purposeful action. The 2SM conceptualizes a person's ability to pursue free, effective, and purposeful action within the game environment as agency. In other words, people who are interacting with a game gain or maintain agency when they feel that they can direct the game toward their personal goals. Conversely, when a person does not feel that he or she can affect the game in a way that feels meaningful, we say that he or she has lost agency. One of the two stances—the Player stance—is specifically oriented toward maintaining this sense of agency, by collecting and organizing practical knowledge on how personal goals may be achieved. Thus, the desire to continue engagement, maintain agency, and advance personal goals provides the impetus for continued play and learning.

3.3.4 Putting It All Together: The Proposed 2SM in Action

The constructs described thus far are, for the most part, closely related to well-established concepts in the research literature. What follows is a description of the hypothetical mechanism by which these constructs interact during game play, and how these interactions may influence thinking and learning within educational games. This description is speculative; we offer it to illustrate how the individual components of the 2SM fit together into a more coherent whole. Many of the specifics of this description are supported by the literature; some other specifics will require future research.

Let us assume that a person begins play with a small initial store of motivation to engage but no knowledge of the game's goals or its interactive model. The person's first instinct is to become situated within the environment, find the useful interfaces, and test the affordances of the environment with tentative actions. At this stage, previous experience playing similar games becomes important; if the person recognizes this particular game as a variant of a genre he or she has played before the person may cue all of his or her existing knowledge as part of process of becoming situated in the game environment. Sometime during this process, and depending on the game structure, a goal will be suggested or will suggest itself to the player's thinking, immediately triggering a self-query, "how do I achieve this goal?" The self-query shifts the person toward a Learning stance, and a second-order model is constructed in response to the query. This second-order model may be partial, inaccurate, or inconsistent, but at this stage, its only requirement is that it suggests one or more steps that might bring the state of the interactive model closer to the goal state. These steps are rendered as heuristics ("When this, Do that") and relayed to the interactive model through whatever controls or interfaces the game allows.

The interactive model processes the player's actions and outputs the appropriate response. The person is now in a position to evaluate the effect of the executed steps in terms of their effectiveness at modifying the state of the interactive model toward the goal state. Actions that prove effective are reinforced and actions that have a negative effect are rephrased as avoidance steps ("Don't do that"). Actions of negligible, ambiguous, or indeterminable effectiveness are discarded. With repeated reinforcement, the player will begin linking the effectiveness of an action with the circumstances that are present whenever that action is invoked. Thus, effective rules, both execution and avoidance, are matched to the context cues from the virtual environment and stored as heuristics, i.e., "If this, do that." These heuristics are easy to remember, quick to access, and require nearly no cognitive effort to execute.

Whenever players find themselves in a situation that is covered by one of their heuristics, they will in most cases attend to that heuristic. In other cases, i.e., when the current conditions cannot be matched with the conditional part of any heuristic, players shift to a Learning stance and reexamine the second-order model and use it to find new possible actions. In this manner, a person's second-order model can evolve cumulatively and iteratively, but only if (a) the person is prompted by the absence of productive heuristics and (b) the person reconsiders the second-order model for the purpose of generating new heuristics. If the player is never without a heuristic to apply, the second-order model can be disregarded as the person simply defaults to the available heuristics.

By the end of the interaction, a person has gathered three forms of knowledge about the game: (1) a collection of observations about the particular conditions that the game presents or can present, (2) a set of heuristics, i.e., rules of action whose activation criteria match up to these conditions, and (3) a second-order model, or hybrid mental model, comprising a network of entities and causal relationships involved in the interactive model and working theories of how these entities and relationships influence the game's structure of experience.

The person can reflect and communicate differently about these three forms of knowledge. Observations and depictions of states and conditions within the game can be readily described verbally, with the aid of physical gestures or other visual aids. Once a semantic domain, that is, a shared mapping of meanings to symbols (see Gee, 2007) to describe the game has been established, communicating heuristics is equally simple, either verbally as an if-then statement, or through a demonstration. The person's second-order model, however, is much more difficult to communicate—language has fewer tools for expressing networked causal relationships. More often than not, these relationships are rendered piecemeal as if-then statements, in a manner resembling p-prims (diSessa, 1993), e.g., "Pulling harder on the slingshot makes the Angry Bird fly further." Second-order models are heavily dependent on the person's specific play experience and trajectory, so that two different people playing the same game might develop two distinct mental depictions of the game's structures. Conversely, people playing the same game (assuming they are equally focused on optimal performance) will converge on similar heuristics. Also, with increasing expertise and experience, and the development of useful heuristics to guide play, it is likely that second-order models are no longer cued at all. These factors combine to make heuristics far more available to people, both individually and as a community, than second-order models. This may also potentially incentivize the learning of heuristics during play (rather than the characteristics of the interactive model) since heuristics are more visible and accessible hallmarks of playing skill.

3.4 Implications for Design and Learning

Thus far, we have conceptualized how learning from games happens from a 2SM perspective and presented evidence from diverse research programs that support the constructs and processes here proposed. We will now discuss the implications of 2SM as it impacts the design of learning environments and the range of learning outcomes that are possible.

3.4.1 Goals for Design

From a 2SM perspective, the success of a game as a pedagogical tool depends mainly on two factors: (1) whether or not the student's second-order model remains accessible after it loses its value as a tool to guide effective play, and (2) whether the second-order model and heuristics that the student generates during play are useful in the target domain. The goals of the designer are therefore (1) to build enough support, feedback, and reinforcement into the learning environment so that the learner second-order model is strengthened throughout play, and (2) to structure game mechanics such that the second-order model and likely heuristics for optimal play support understanding or problem-solving in the target domain. The negotiation of these priorities might shift according to how a particular domain's notion of expert knowledge balances between abstract yet flexible understanding of a system's workings and quick execution of procedures that are known to be effective. In the case of the former, the second-order model is the privileged form of knowledge; the goal becomes to enable the second-order model to enter the student's long-term memory as the kind of context-free causal/relational cognitive structure that more resembles what some scholars view as expert knowledge (cf. Chi, Feltovich, & Glaser, 1981). On the other hand, if heuristics are the preferred mode of knowledge for a particular domain, then it is likely the development of expertise via games more closely resembles skill acquisition (cf. Anderson, 1987).

3.4.2 Alignment Between External Model, Interactive Model, and Second-Order Model

An ongoing challenge in game-based learning research is the matter of "alignment," namely, ensuring that what students actually learn from an educational game coincides with both the designers' intent and an externally validated curriculum (Kebritchi, 2010; Squire & Jenkins, 2003). Mental model-based explanations of student learning generally frame this alignment as a matter of creating accurate facsimiles of the formal abstractions of scientific phenomena, encoding them into the interactive model, and providing the necessary tools and scaffolds so students can best "make sense of" the interactive model. Accurate interactive models, along with useful scaffolds to make the model more clear and apparent, are hypothesized to result in more effective mental models.

The 2SM shifts the issue of alignment from accuracy in the sense of fidelity of representation to accuracy in the sense of proximity between the task demands of the game and those of the curriculum topic of interest (for a similar treatment of this idea, see also Holbert & Wilensky, 2014). A good example of this latter form of proximity is *Dragon Box*, a puzzle game intended to help students learn concepts of algebra. In *Dragon Box*, players manipulate cards on a two-sided game board with the goal of clearing the board of cards, leaving only a "dragon" card and a "box" to place it in. In algebraic terms, the board represents a mathematical equality, the cards represent coefficients, and the box represents the unknown variable.

From the 2SM perspective, *Dragon Box* succeeds, at least in part, because the second-order model that students form while playing the game is closely associated to the form that algebra problems tend to take, e.g., isolating the variable, or balancing the equation. Not only are the steps and processes similar in both the game and the real-world application, but so are the goals. This similarity in goal structure is rather unique. Researchers have noted that games could portray not only the knowledge base of science and the material methods of science inquiry, but also its purposes, priorities, and objectives, e.g., Shaffer, Squire, Halverson, and Gee (2005) and Barab, Gresalfi, and Ingram-Goble (2010). One example of this enriched portrayal of science can be found in the *Quest Atlantis* games, or in the *FoldIt* protein research simulation/puzzle game. However, the disconnection between the goal structure of *Quest Atlantis* and that of the typical science unit is inevitably felt by students, and thus the mental structures that students form to successfully navigate both forms of learning are perhaps not interoperable. *FoldIt* and *Dragon Box*, on the other hand, hew so closely to the goal structures of the discipline that alignment is tight; students may see the game and the respective disciplinary practices as being integrated procedural knowledge. Thus, from the 2SM perspective, *Dragon Box* is a prime example of a game that successfully manages the "alignment" challenge and structures students' thinking in a manner closely tied to the learning domain targeted by the game.

3.4.3 Helping Students Form Robust Mental Models

Helping students form robust mental models that can be applied across contexts has also historically been a strong focus of inquiry. In this sense, a "robust model" is one that brooks no inconsistency and helps eliminate confusion. de Kleer and Brown (1980) define mental model robustness as the ability of the model to correctly answer questions about the structures it describes. A robust mental model manifests itself as a thorough, qualitative understanding of the mechanism of interest, often with no recourse to formal descriptions (such as equations, technical nomenclature, diagrams) and not limited to describing or explaining any one particular example. It may be fairly said that the development of these type of mental models is the main goal of science education (Gilbert & Boulter, 2012).

Educational games that aim to develop robust mental models are thus a potentially significant avenue of inquiry. Research from the past three decades provides substantial insight into the design of games to support students' ability to engage with them, understand them, and use them as tools for thinking. In fact, this goal has arguably been the overall design imperative in research on educational games, in line with existing principles of instructional design and multimedia learning (e.g., Clark & Mayer, 2011). Yet behind these principles lies the baseline assumption that once learners form their mental models, these models remain available to students at some later time. The 2SM framework problematizes that assumption by suggesting that second-order models are System 2 processes, and like all System 2 processes, they are preempted by System 1 processes such as rules and heuristics whenever these are available (Schwartz & Black, 1996). In fact, given that System 1 mechanisms are the preferred mechanism for everyday thought, it is perhaps incidental that students retain any knowledge at all resembling a mental model that is useful in new contexts.

Thus, the challenge for designers of educational games is to help students store and recall their second-order models, which may run counter to the students' own cognitive biases against using them. The question then becomes, how can designers disrupt a person's natural tendency to discard second-order models? A person is most likely to stop attending to his or her second-order model (a) once effective rules or heuristics have been derived from the second-order model and (b) once the heuristics cover all possible situations the player cares to affect. With regard to the first condition, it is probably not feasible to prevent students from forming heuristics.

People have strong incentive to maintain agency, and they express this agency by directing the interactive model toward a desired state. Effective control-oriented rules are therefore constantly being created, selectively matched to available data, and evaluated for effect (in a matter reminiscent of the description of production systems in Neches, Langley, & Klahr, 1987). These cognitive processes are associative and prone to automatization. It is therefore unlikely that the designer can interrupt them without damaging people's agency and the ability to continue to engage.

The second approach to preserving second-order models is potentially more promising. If a game is designed in such a way that it is constantly offering new goals as well as elements and constraints, then at no time do a person's heuristics provide good-enough play actions for all possible situations. On the contrary, the person must return to his or her second-order model (from Player stance to Learner stance) to revise and refine it to include the new structures. If the person does not expand his or her second-order model, then the second-order model will lose its usefulness at predicting states in the interactive model and evaluating the effectiveness of the potential actions. In this situation, the person cannot retain agency, and his or her ability to engage effectively quickly decays. Thus, whenever the game introduces new elements and these new elements are different enough that the person cannot cope with them using existing heuristics, the person shifts from Player stance back into the Learner stance, the second-order model is re-invoked, and the process of modeling begins again (so person can eventually shift back to Player stance). If this chain of events happens frequently enough, the person will reinforce the mental model, rather than the set of heuristics. This may, in turn, result in greater availability of the mental model for problem-solving in other contexts.

Dragon Box appears to effectively transition users between Player stance and Learner stance with a very extensive and gradual level progression. *Dragon Box* contains 200 levels, each building on the prior levels, and almost all of them add a new quirk, wrinkle, or complexity. As new symbols and new rules are introduced, students must constantly adjust their second-order models to keep pace. This prevents students from settling on effort-reducing strategies such as heuristics; however, the heuristics that do form are closely aligned the tasks and subtasks found in algebra problems in any case.

A similar example of the effect of constantly evolving challenges can be found in the recreational game *Dwarf Fortress*, a construction and management simulation akin to *SimCity* or *Railroad Tycoon*. What makes *Dwarf Fortress* different is the depth and detail of its modeled world, whose principles operate with a regularity and complexity far beyond most digital games. The game includes systems to simulate basic economic activities like farming, fishing, hunting, and a broad variety of crafts, such as smithing, masonry, and brewing. Each of these activities is supported by simulations of resource growth and propagation (i.e., seeds that grow into plants that bear fruit, fish and wild game that reproduce, and predators that compete with the dwarves for the same food resources), and the behavior and interactions of materials (e.g., wood burns, iron melts, bones decompose, and water that flows into magma produces obsidian and steam).

The complexity and depth of *Dwarf Fortress*' interactive model means that the learning curve for the game is uncommonly steep. Furthermore, there are specific in-game events that, when game conditions are met, trigger an explosion of systemic complexity. Thus, players remain constantly off-balance: just when the current difficulty fades as the challenges are mastered, new and more complex goals appear, exposing new functionalities of the interactive model. From the 2SM perspective, second-order models are constantly undergoing revision and are never quite discarded, and conversely, the heuristics that people form are (at best) very general guidelines to smooth play, because no general-purpose or always-applicable rules are possible (Martinez-Garza, 2015).

Dwarf Fortress exposes a tension, however, between the equally desirable goals of (a) proficiency with the concepts, entities, and relationships of the interactive model and (b) learners' sense of self-efficacy and motivation. There is something fundamentally off-putting about complex games like *Dwarf Fortress*. The sense of disorientation that they produce, of not knowing exactly all that is going on, is arguably not an optimal starting point for learning. Educational games depend largely on their motivational affordances for their buy-in. Teachers and educators are receptive to games because students tend to find them engaging. It thus seems counterproductive to make games complex and disorienting. On the other hand, the depth and responsiveness of the interactive model encased in *Dwarf Fortress* is closest to the letter and spirit of the justification for using interactive models in the first place, viz. to allow users to investigate the causes of phenomena, and explore the implications of manipulating certain parameters. This tension is perhaps not one that can be resolved, but it can be studied and negotiated through skillful design.

3.4.4 Attending to Social Texture

Designing educational games from the perspective of the 2SM also requires the designer to attend to the social texture of the context in which the game is played. The two stances are epistemological constructs, and as such, refer mainly to private cognition. However, the boundaries of these stances are not impermeable, and can inform and be informed by social interaction. Our conjecture is that most such interactions will center on the circulation of heuristics. Heuristics, as described in the 2SM, are a form of knowledge that is easy to transmit, easy to decode, easy to remember, and highly portable. As such, a person's heuristics can become social capital and commodities among player communities. Effective heuristics are prized by players because these heuristics are often the difference between progress and frustration, between satisfying and unsatisfying play. The person who describes his or her heuristics to others gains social currency, a form of prestige that comes from being the person who "figured it out." This effect can be observed both in classrooms, where ad hoc cooperative play and helping behaviors are common, as well as in distributed online spaces, where game knowledge is freely shared in a form of *potlatch* or gift economy. The skillful *use* and application of this knowledge is also encouraged when it directly improves a person's level of play since this improvement is a source of prestige (as shown by Schrader & McCreery, 2008).

The heuristics component of the 2SM supports transactivity in this regard. While second-order models are unwieldy to communicate and share, heuristics are easy to express both in written and spoken form, and the phenomenological regularity of the common game experience means that heuristics are difficult to misunderstand or misapply. Heuristics thus support online social interactions around the game, often in the form of prescriptions for expert play (e.g., FAQs,[2] "walkthroughs," "cheats," or tips) and provide impetus for the affinity spaces that organize these interactions around learning and mastery (i.e., "big-G games," as discussed by Gee, 2008). These transactions are framed largely in terms of how to play games optimally, with comparatively little emphasis on formal analysis of how the game works.

Cognitive explanations of game-based learning often rely on interpreting the interactions of the learner with the learning environment rather than taking a broader situated view encompassing collaboration and community. This is not an insurmountable limitation in the case of games that are structured around one-to-one user-to-computer interactions. Yet educational games are frequently used in classrooms without clearly defined user boundaries; it is very difficult to determine the degree to which a mental analogue arises from individual cognitive effort as opposed to participation alongside other students in a collaborative enterprise. In fact, even in games designed for solo play, ad hoc collaborations between students are more likely the norm and not the exception (e.g., the student-directed collaborative task pursuit in single-player games described by Sharritt (2008) and Nilsson and Jakobsson (2011). Other naturalistic settings, such as massively multiplayer online games, also share this very thin border between private and shared cognition.

The 2SM provides a plausible structure for these collaborations. Essentially, students in classrooms use a sort of *distributed* Learner stance. When individuals induce effective rules through the processes described above, they can make them available to others as needed. If people become stuck in the game, they have the option of consulting their peers instead of querying their own second-order models. If a rule of effective action is available, more often than not a more proficient player will share it, although help-seeking and help-giving behavior based on the exchange of heuristics is likely mediated by the norms and sociocultural practices that operate in that particular classroom. If a rule of effective action is not available or forthcoming, then students can continue playing on their own or collaborate in a shared Learner stance until a rule is found. These ad hoc collaborations are made possible by the portability and context-unbound nature of heuristics. If heuristics did not have these qualities, help-seeking and help-giving would involve higher cost in time and effort. Ad hoc collaborations would thus be far rarer.

[2] For "frequently asked questions," a genre of guide document in which a community of more expert players collect information for the benefit of more novice players in an effort to limit redundancy. This genre is described extensively by Gee (2003), along with its print analogue, the "strategy guide." The fact that these documents exist indicates that different players have consistent enough experiences so that many questions become "frequently asked."

3.4.5 Connecting Educational and Leisure Games

One final issue concerns the applicability of the 2SM (or any learning theory) to both educational and leisure games. It may be argued that games for learning require different theories than leisure games in order to account for the added demands of learning. Yet to a large extent, the 2SM treats educational and leisure games as one and the same. The fact that certain games are conceptualized as helping to teach specific content is more an artifact of their design and intended application than any departure from the medium as a whole (see Gee, 2003). From a 2SM perspective, there are only two main differences. First, the elective versus compulsory nature of out-of-school versus in-school gaming probably has some bearing on the resources available to the Player stance. Second, classrooms generally present lower barriers to collaboration compared to gaming "in the wild" due to the direct physical colocation of the participants, emphasizing the importance of attending to the sociocognitive structure of the game-playing community. In other respects, the 2SM treats educational games exactly like leisure games; they use the same structures of experience, the same design language, the same technologies, the same genre conventions, and the same representational practices. Thus, we propose that we might prefer a unitary theory like the 2SM, one that allows for taxonomies and contexts yet helps explain the thinking and learning of all players of all kinds of games.

3.5 Conclusions

This chapter began by focusing on a perplexing duality; players of digital games appear to both act automatically and reflect deeply. We have proposed that this duality, as problematic as it might seem if we assume that learning only happens during moments of analysis and reflection, is perhaps not specific to digital games, but rather a feature of human cognition in general. As such, there is significant and persuasive scholarship that demonstrates how this duality in our thinking and learning capacities works, and we have endeavored to synthesize and summarize this research here.

The Two Stance Framework represents a first attempt to reify this general theory of cognition explicitly into the realm of game-based learning. The 2SM, as we have argued here, has a number of promising features that shed some light on the persistent challenge of designing a game that helps students learn in such a way that their improved performance in-game has some bearing on their proficiency out-of-game. Among these features are (a) improved explanatory power regarding intrapersonal variation in learning from games; (b) more complete theory regarding individual needs, goals, and agency; (c) a more extensive account of collaboration and community; and (d) improved perspective on knowledge-rich interactions in online affinity spaces. These affordances harmonize well with existing theories of learning. Clearly, there are issues regarding game-based learning that the 2SM can probably not address. In these cases, we have tried to limit our claims by deferring to more applicable theory in matters of scope (as in the initial sections of this chapter) or by indicating where the 2SM might share a point of contact and focus with general theories of learning, as we have done in the latter part of our argument.

While the 2SM is conjectural, and therefore destined to undergo revision and refinement, the research literature provides promising indicia that support its general premises. The work that remains is to craft specific investigations to demonstrate the 2SM empirically. Meeting this challenge will require a mix of observational and quantitative methodologies including sophisticated pattern-finding analytics that support linking game behaviors to epistemological stances to learning outcomes. Fortunately, this combined methodological approach can leverage a burgeoning foundation of tools in the field of game-based learning, as scholars recognize and seek to account for the richness and complexity of game-based phenomena. Enriched multiple-perspective research strategies, supported by existing theory as well as new frameworks like the 2SM, promise to support more sophisticated student understanding, help learners build more powerful identities, and advance our understanding of the rapidly evolving world of digital gaming.

Acknowledgements: The research reported here was supported by the National Science Foundation through grant 1119290 and the Institute of Education Sciences, U.S. Department of Education, through grant R305A110782. The opinions expressed are those of the authors and do not represent views of the Institute, the U.S. Department of Education, or the National Science Foundation.

References

Anderson, J. R. (1987). Skill acquisition: Compilation of weak-method problem situations. *Psychological Review, 94*(2), 192–210.

Anderson, J., & Barnett, M. (2011). Using video games to support pre-service elementary teachers learning of basic physics principles. *Journal of Science Education and Technology, 20*(4), 347–362.

Apostel, L. (1961). Towards the formal study of models in the non-formal sciences. In *The concept and the role of the model in mathematics and natural and social sciences* (pp. 1–37). Dordrecht: Springer. Retrieved from http://link.springer.com/chapter/10.1007/978-94-010-3667-2_1

Barab, S. A., Gresalfi, M. S., & Ingram-Goble, A. (2010). Transformational play: Using games to position person, content, and context. *Educational Researcher, 39*(7), 525–536.

Bartle, R. (1996). Hearts, clubs, diamonds, spades: Players who suit MUDs. *Journal of MUD Research, 1*(1), 19.

Bekebrede, G., & Mayer, I. (2006). Build your seaport in a game and learn about complex systems. *Journal of Design Research, 5*(2), 273–298. doi:10.1504/JDR.2006.011366.

Chaiken, S. (1980). Heuristic versus systematic information processing and the use of source versus message cues in persuasion. *Journal of Personality and Social Psychology, 39*(5), 752–766. doi:10.1037/0022-3514.39.5.752.

Chater, N., & Oaksford, M. (1998). *Rational models of cognition*. Oxford: Oxford University Press.

Chi, M. T., Feltovich, P. J., & Glaser, R. (1981). Categorization and representation of physics problems by experts and novices. *Cognitive Science, 5*(2), 121–152.

Chi, M. T., Glaser, R., & Rees, E. (1981). *Expertise in problem solving*. Pittsburgh, PA: Learning Research and Development Center, University of Pittsburgh.

Clark, D. B., & Jorde, D. (2004). Helping students revise disruptive experientially supported ideas about thermodynamics: Computer visualizations and tactile models. *Journal of Research in Science Teaching, 41*(1), 1–23.

Clark, R. C., & Mayer, R. E. (2011). *E-learning and the science of instruction: Proven guidelines for consumers and designers of multimedia learning* (3rd rev. ed.). Chichester, UK: Jossey Bass Wiley.

Clark, D., Nelson, B., Sengupta, P., & D'Angelo, C. (2009, October). Rethinking science learning through digital games and simulations: Genres, examples, and evidence. In *Learning science: Computer games, simulations, and education workshop sponsored by the National Academy of Sciences*, Washington, DC.

Clement, J. (2000). Model based learning as a key research area for science education. *International Journal of Science Education, 22*(9), 1041–1053.

de Kleer, J., & Brown, J. S. (1980). Mental models of physical mechanisms and their acquisition. In J. R. Anderson (Ed.), *Cognitive skills and their acquisition* (pp. 285–309). Hillsdale, NJ: Erlbaum.

Deci, E. L., & Ryan, R. M. (1985). *Intrinsic motivation and self-determination in human behavior*. New York: Plenum Publishing Company.

Detterman, D. (1993). The case for prosecution: Transfer as an epiphenomenon. In D. Detterman & R. Sternberg (Eds.), *Transfer on trial: Intelligence, cognition and instruction* (pp. 1–24). Norwood, NJ: Alex.

diSessa, A. A. (1993). Toward an epistemology of physics. *Cognition and Instruction, 10*(2–3), 105–225.

Doyle, J. K., & Ford, D. N. (1998). Mental model concepts for system dynamics research. *System Dynamics Review, 14*(1), 3–29.

Elby, A., & Hammer, D. (2001). On the substance of a sophisticated epistemology. *Science Education, 85*(5), 554–567.

Epstein, S. (1994). Integration of the cognitive and the psychodynamic unconscious. *American Psychologist, 49*(8), 709–724. doi:10.1037/0003-066X.49.8.709.

Evans, J. S. B. T. (2003). In two minds: Dual-process accounts of reasoning. *Trends in Cognitive Sciences, 7*(10), 454–459. doi:10.1016/j.tics.2003.08.012.

Evans, J. S. B. T. (2006). The heuristic-analytic theory of reasoning: Extension and evaluation. *Psychonomic Bulletin & Review, 13*(3), 378–395. doi:10.3758/BF03193858.

Evans, J. S. B. T. (2008). Dual-processing accounts of reasoning, judgment, and social cognition. *Annual Review of Psychology, 59*(1), 255–278. doi:10.1146/annurev.psych.59.103006.093629.

Evans, J. S. B. T., & Stanovich, K. E. (2013). Dual-process theories of higher cognition advancing the debate. *Perspectives on Psychological Science, 8*(3), 223–241. doi:10.1177/1745691612460685.

Fiske, S. T., & Taylor, S. E. (1991). *Social cognition* (2nd ed.). New York, NY: McGraw-Hill.

Gee, J. P. (2003). *What Video Games Have to Teach Us about Learning and Literacy*. New York: Palgrave Macmillan.

Gee, J. P. (2007). *What video games have to teach us about learning and literacy*. Second edition: Revised and updated edition (2nd ed.). New York, NY: Palgrave Macmillan.

Gee, J. P. (2008). Learning and games. In K. Salen (Ed.), *The ecology of games: Connecting youth, games and learning* (pp. 21–40). Cambridge, MA: MIT Press.

Gigerenzer, G., & Goldstein, D. G. (1996). Reasoning the fast and frugal way: models of bounded rationality. *Psychological Review, 103*(4), 650.

Gigerenzer, G., Hoffrage, U., & Kleinbölting, H. (1991). Probabilistic mental models: A Brunswikian theory of confidence. *Psychological Review, 98*, 506–528.

Gigerenzer, G., & Selten, R. (Eds.). (2001). *Bounded rationality: The adaptive toolbox*. Cambridge, MA: MIT Press.

Gilbert, J. K., & Boulter, C. (Eds.). (2012). *Developing models in science education*. Berlin: Springer Science & Business Media.

Hammer, D., & Elby, A. (2003). Tapping epistemological resources for learning physics. *The Journal of the Learning Sciences, 12*(1), 53–90.

Hammer, D., Elby, A., Scherr, R. E., & Redish, E. F. (2005). Resources, framing, and transfer. In J. Mestre (Ed.), *Transfer of learning from a modern multidisciplinary perspective* (pp. 89–120).

Hasher, L., & Zacks, R. T. (1979). Automatic and effortful processes in memory. *Journal of Experimental Psychology: General, 108*, 356–388.

Holbert, N. R., & Wilensky, U. (2014). Constructible authentic representations: Designing video games that enable players to utilize knowledge developed in-game to reason about science. *Technology, Knowledge and Learning, 19*(1–2), 53–79.

Johnson-Laird, P. N. (1983). *Mental models: Towards a cognitive science of language, inference, and consciousness* (Vol. 6). Cambridge, MA: Harvard University Press.

Jones, M. G., Minogue, J., Tretter, T. R., Negishi, A., & Taylor, R. (2006). Haptic augmentation of science instruction: Does touch matter? *Science Education, 90*(1), 111–123.

Kahneman, D. (2003). Maps of bounded rationality: Psychology for behavioral economics. *The American Economic Review, 93*(5), 1449–1475.

Kebritchi, M. (2010). Factors affecting teachers' adoption of educational computer games: A case study. *British Journal of Educational Technology*, 41(2), 256–270.

Lehrer, R., & Schauble, L. (2005). Cultivating model-based reasoning in science education. In R. K. Sawyer (Ed.), *The Cambridge handbook of the learning sciences* (pp. 371–388). Cambridge, UK: Cambridge University Press. doi:10.1017/CBO9780511816833.023.

Li, Q. (2010). Digital game building: Learning in a participatory culture. *Educational Research, 52*(4), 427–443.

Marino, M. T., Basham, J. D., & Beecher, C. C. (2011). Using video games as an alternative science assessment for students with disabilities and at-risk learners. *Science Scope, 34*(5), 36–41.

Martinez-Garza, M. M. (2015). Examining epistemic practices of the community of players of *Dwarf Fortress*: "For !!SCIENCE!!". *International Journal of Gaming and Computer-Mediated Simulations (IJGCMS), 7*(2), 46–67.

Mayer, R. E. (2005). Cognitive theory of multimedia learning. In R. E. Mayer (Ed.), *The Cambridge handbook of multimedia learning* (pp. 31–48). Cambridge, UK: Cambridge University Press.

Mayer, R. E., & Wittrock, M. C. (1996). Problem-solving transfer. In D. C. Berliner & R. C. Calfee (Eds.), *Handbook of educational psychology* (pp. 47–62). Mahwah, NJ: Lawrence Erlbaum Associates.

McNamara, D. S., & Shapiro, A. M. (2005). Multimedia and hypermedia solutions for promoting metacognitive engagement, coherence, and learning. *Journal of Educational Computing Research, 33*(1), 1–29.

Moreno, R., & Mayer, R. E. (2000). Engaging students in active learning: The case for personalized multimedia messages. *Journal of Educational Psychology, 92*(4), 724–733.

Moreno, R., & Mayer, R. E. (2005). Role of guidance, reflection, and interactivity in an agent-based multimedia game. *Journal of Educational Psychology, 97*(1), 117–128.

Neches, R., Langley, P., & Klahr, D. (1987). Learning, development, and production systems. In D. Klahr, P. Langley, & R. Neches (Eds.), *Production system models of learning and development* (pp. 1–53). Cambridge, MA: The MIT Press.

Nilsson, E. M., & Jakobsson, A. (2011). Simulated sustainable societies: Students' reflections on creating future cities in computer games. *Journal of Science Education and Technology, 20*(1), 33–50.

Nisbett, R. E., Peng, K., Choi, I., & Norenzayan, A. (2001). Culture and systems of thought: Holistic versus analytic cognition. *Psychological Review, 108*(2), 291–310. doi:10.1037/0033-295X.108.2.291.

Perkins, K., Adams, W., Dubson, M., Finkelstein, N., Reid, S., Wieman, C., et al. (2006). PhET: Interactive simulations for teaching and learning physics. *The Physics Teacher, 44*(1), 18–23.

Rosenbaum, E., Klopfer, E., & Perry, J. (2007). On location learning: Authentic applied science with networked augmented realities. *Journal of Science Education and Technology, 16*(1), 31–45. doi:10.1007/s10956-006-9036-0.

Rouse, W. B., & Morris, N. M. (1986). On looking into the black box: Prospects and limits in the search for mental models. *Psychological Bulletin, 100*(3), 349–363. doi:10.1037/0033-2909.100.3.349.

Ryan, R. M., Rigby, C. S., & Przybylski, A. (2006). The motivational pull of video games: A self-determination theory approach. *Motivation and Emotion, 30*(4), 344–360.

Schrader, P. G., & McCreery, M. (2008). The acquisition of skill and expertise in massively multiplayer online games. *Educational Technology Research and Development, 56*(5–6), 557–574.

Schwartz, D. L., & Black, J. B. (1996). Shuttling between depictive models and abstract rules: Induction and fallback. *Cognitive Science, 20*(4), 457–497.
Shaffer, D. W., Squire, K. R., Halverson, R., & Gee, J. P. (2005). Video games and the future of learning. *Phi Delta Kappan, 87*(2), 104–111.
Sharritt, M. J. (2008). Forms of learning in collaborative video game play. *Research and Practice in Technology Enhanced Learning, 3*(2), 97–138.
Sherry, J. L., Lucas, K., Greenberg, B. S., & Lachlan, K. (2006). Video game uses and gratifications as predictors of use and game preference. In P. Vorderer & J. Bryant (Eds.), *Playing video games: Motives, responses, and consequences* (pp. 213–224). Mahwah, NJ: Lawrence Erlbaum Associates.
Simon, H. A. (1956). Rational choice and the structure of the environment. *Psychological Review, 63*(2), 129–138. doi:10.1037/h0042769.
Sloman, S. A. (1996). The empirical case for two systems of reasoning. *Psychological Bulletin, 119*(1), 3–22. doi:10.1037/0033-2909.119.1.3.
Smith, E. E., Langston, C., & Nisbett, R. E. (1992). The case for rules in reasoning. *Cognitive Science, 16*(1), 1–40. doi:10.1207/s15516709cog1601_1.
Squire, K. D., & Jenkins, H. (2003). Harnessing the power of games in education. *Insight, 3*(1), 5–33.
Stanovich, K. E. (1999). *Who is rational? Studies of individual differences in reasoning*. Mahwah, NJ: Lawrence Erlbaum Associates.
Stenning, K., & Van Lambalgen, M. (2008). *Human reasoning and cognitive science*. Cambridge, MA: MIT Press.
Taylor, M. J., Pountney, D. C., & Baskett, M. (2008). Using animation to support the teaching of computer game development techniques. *Computers & Education, 50*(4), 1258–1268. doi:10.1016/j.compedu.2006.12.006.
Veldhuyzen, W., & Stassen, H. G. (1977). The internal model concept: An application to modeling human control of large ships. *Human Factors: The Journal of the Human Factors and Ergonomics Society, 19*(4), 367–380.
Vosniadou, S., & Brewer, W. F. (1994). Mental models of the day/night cycle. *Cognitive science, 18*(1), 123–183.
Yee, N. (2006). Motivations for play in online games. *CyberPsychology & Behavior, 9*(6), 772–775.

Chapter 4
Assessment and Adaptation in Games

Valerie Shute, Fengfeng Ke, and Lubin Wang

Abstract Digital games are very popular in modern culture. We have been examining ways to leverage these engaging environments to assess and support important student competencies, especially those that are not optimally measured by traditional assessment formats. In this chapter, we describe a particular approach for assessing and supporting student learning in game environments—stealth assessment—that entails unobtrusively embedding assessments directly and invisibly into the gaming environment. Results of the assessment can be used for adaptation in the form of scaffolding, hints, and providing appropriately challenging levels. We delineate the main steps of game-based stealth assessment and illustrate the implementation of these steps via two cases. The first case focuses on developing stealth assessment for problem-solving skills in an existing game. The second case describes the integration of game and assessment design throughout game development, and the assessment and support of mathematical knowledge and skills. Both cases illustrate the applicability of data-driven, performance-based assessment in an interactive game as the basis for adaptation and for use in formal and informal contexts.

Keywords Stealth assessment • Adaptation • Bayesian networks

4.1 Introduction

According to "*2015 Essential Facts About the Computer and Video Game Industry*" published by Entertainment Software Association, over 150 million Americans play video games and 42 % play regularly for at least 3 h per week. The popularity of video games has drawn researchers' attention in the exploration of the possibility of using video games to enhance knowledge, skills, and other personal attributes. The idea of using games for serious purposes other than entertainment is called game-based learning. Advocates of game-based learning argue that well-designed

V. Shute (✉) • F. Ke • L. Wang
Educational Psychology and Learning Systems Department,
Florida State University, 3205G Stone Building, 1114 West Call Street,
Tallahassee, FL 32306-4453, USA
e-mail: Vshute@fsu.edu

P. Wouters, H. van Oostendorp (eds.), *Instructional Techniques to Facilitate Learning and Motivation of Serious Games*, Advances in Game-Based Learning, DOI 10.1007/978-3-319-39298-1_4

video games represent solid learning principles such as providing ongoing feedback, interactivity, meaningful and engaging contexts, and adaptive challenges within the zone of proximal development (Bransford, Brown, & Cocking, 2000; Gee, 2003; Shute, 2008; Vygotsky, 1978). A fair amount of research shows that game-based learning is at least as effective as nongame conditions, such as classroom contexts (e.g., Barab, Gresalfi, & Ingram-Goble, 2010; Clark, Tanner-Smith, & Killingsworth, 2014; Sitzmann, 2011; Wouters, van Nimwegen, van Oostendorp, & van der Spek, 2013).

Researchers are also beginning to realize that games can serve as effective assessments (e.g., DiCerbo & Behrens, 2012; Shute, Leighton, Jang, & Chu, 2016; Shute & Ventura, 2013). That is, while players interact with the game environment, the game engine monitors and collects information about players' performances and provides feedback to players in the form of in-game scores or the avatar's progress in the game. This is basically the same as what educational assessment does, i.e., making inferences about students' knowledge and skills by observing what students say, do, and produce in a given context (Mislevy, Steinberg, & Almond, 2003). In addition, when game-based assessment is designed following a principled assessment design framework such as evidence-centered design (ECD; Mislevy et al., 2003) or cognitive design system (CDS; Embretson, 1998), the assessment is likely to have high validity and reliability.

Game-based assessment is essentially performance-based assessment. Performance-based assessment refers to tasks that require students to demonstrate their knowledge and skills by working through a task (Flynn, 2008; Madaus & O'Dwyer, 1999). Rather than a simple test of one's ability to recall or recognize information, or supply self-reported information, performance-based assessment provides students with the opportunity to show their understanding and apply knowledge in meaningful settings (Stecher, 2010). Scholars generally support the use of performance-based assessment to measure and support twenty-first-century skills (e.g., problem solving, creativity, collaboration; Partnership for the 21st Century 2015) over conventional types of assessment such as multiple-choice questions or filling in the blanks (see Shute et al., 2016). However, there are a few challenges associated with the design and implementation of performance-based assessments. Some of the more difficult challenges include: (a) designing contexts that will fully elicit the competencies to be measured, (b) modeling the multidimensionality of constructs to be measured, (c) ensuring the validity and reliability (consistency) of the tasks, (d) providing appropriate feedback that is customized to each individual situation, (e) automating the scoring of the various tasks, (f) accumulating the evidence across all task performances, and (g) reducing the development costs of performance-based assessments compared to traditional tests. Our premise in this chapter is that stealth assessment (see Shute, 2011) coupled with ECD provides a viable solution to these challenges.

In addition to serving as assessment vehicles, games can help to support learning and motivation. That is, people who want to excel at something spend countless hours making intellectual effort and practicing their craft. But practice can be boring and frustrating, causing some learners to abandon their practice and, hence, learning.

This is where the principles of game design come in—good games can provide an engaging and authentic environment designed to keep practice meaningful and personally relevant. With simulated visualization, authentic problem solving, and instant feedback, computer games can afford a realistic framework for experimentation and situated understanding, and thus act as rich primers for active, motivated learning (Barab, Thomas, Dodge, Carteaux, & Tuzun, 2005; Squire, 2006). Another key feature of well-designed games that can enhance learning and motivation is adaptivity related to providing appropriate and adaptive levels of challenge (see Fullerton, 2014). Gee (2003) has argued that the secret of a good game is not its 3D graphics and other bells and whistles, but its underlying architecture in which each level dances around the outer limits of the player's abilities, seeking at every point to be hard enough to be just doable. Similarly, psychologists (e.g., Vygotsky, 1987) have long argued that the best instruction hovers at the boundary of a student's competence. Flow is another name for this phenomenon. It is a construct first proposed by Csikszentmihalyi (1990, 1997) to describe an optimal experiential state that involves complete immersion in an activity and a deep sense of enjoyment. Flow represents full engagement, which is crucial for deep learning. The essential components of flow include clear and unambiguous goals, challenging yet achievable levels of difficulty, and immediate feedback (Cowley, Charles, Black, & Hickey, 2008; Csikszentmihalyi, 1997). In the game design context, flow theory states that if the player finds a level too difficult, he/she will become frustrated. However, if, as the player continues playing, his/her abilities improve while the challenge level stays the same, he/she will become bored. Therefore, to facilitate a flow state, challenge and ability must be carefully balanced to accomplish this type of adaptivity.

In this chapter, we first review the theoretical foundations of ECD and stealth assessment. In the second section, we discuss how stealth assessment works. After the discussion, we demonstrate the process of creating stealth assessment using ECD via two examples—one past and one current research project—that apply the approach. We then conclude this paper with a brief discussion on implications for future research.

4.2 Literature Review

4.2.1 Evidence-Centered Design

Evidence-centered design (Mislevy et al., 2003) provides a framework for designing and implementing assessments that support arguments about personal competencies via an evidence chain that connects the arguments with task performance. ECD consists of conceptual and computational models that work together. The three major models include the competency model, the evidence model, and the task model.

The *competency model* outlines in a structured fashion the beliefs about personal knowledge, skills, or other learner attributes. The competency model can host unidimensional constructs and, importantly, multidimensional constructs (e.g., problem solving, leadership, and communication skills) as well. The beliefs about learners' competencies in the competency model are updated as new evidence supplied by the evidence model comes in. When competency model variables are instantiated with individual student data, the competency model is often referred to as the student model.

The *task model* identifies the features of selected tasks for learners that will provide evidence about their target competencies. The main function of the task model is to provide observable evidence ,about the unobservable competencies, which is realized via the evidence model.

The *evidence model* serves as the bridge between the competency model and the task model. It transmits evidence elicited by tasks specified by the task model to the competency model by connecting the evidence model variables and competency model variables statistically. Basically, the evidence model contains two parts: (a) evidence rules or rubrics that convert the work products created during the interactions between the learner and the tasks to observable variables that can be scored in the form of "correct/incorrect" or graded responses; and (b) a statistical model that defines the relationships among observable variables and competency model variables, and then aggregates and updates scores across different tasks. The statistical model may be in the form of probabilities based on Bayes theorem or they may be simple cut scores.

4.2.2 Stealth Assessment

Stealth assessment, a specialized implementation of ECD, is a method of embedding assessment into a learning environment (e.g., video games) so that it becomes invisible to the learners being assessed (Shute, 2011). We advocate the use of stealth assessment because of its many advantages. As we mentioned at the beginning of the chapter, there are a number of challenges related to performance-based assessment, but stealth assessment addresses each challenge. Because it is designed to be unobtrusive, stealth assessment frees students from test anxiety commonly associated with traditional tests and thus improves the reliability and validity of the assessment (e.g., DiCerbo & Behrens, 2012; Shute, Hansen, & Almond, 2008). Second, stealth assessment is designed to extract ongoing evidence and update beliefs about students' abilities as they interact with the tasks. This allows assessors to diagnose students' performance and provide timely feedback. As a result, interacting with the learning or gaming environment can support the development of students' competencies as they are being assessed. Third, when stealth assessment is designed following ECD, this allows for the collection of sufficient data about students' target competencies at a fine grain size providing more information about a student's ability compared with conventional types of assessment like multiple-choice formats. Fourth, when stealth assessment is embedded within a well-designed video game, students are fully engaged in the experience, which is conducive to the extraction of true knowledge and skills. Fifth, because scoring in stealth assessment is automated, teachers do not need to spend valuable time calculating scores and grades. Finally, stealth assessment models, once developed and validated, can be reused in other learning or gaming environments with only some adjustments to the particular game indicators.

Recently, we have been creating and testing stealth assessments of various competencies within video games. For instance, we developed and embedded three stealth assessments (running concurrently) of qualitative physics understanding (Shute, Ventura, & Kim, 2013), persistence (Ventura, Shute, & Small, 2014; Ventura, Shute, & Zhao, 2012), and creativity (Kim & Shute, in press) in a homemade game called *Physics Playground*, formerly called *Newton's Playground* (see Shute & Ventura, 2013). We created and tested stealth assessments of problem solving and spatial skills for the commercial game *Portal 2* (Shute, Ventura, & Ke, 2015; Shute & Wang, in press). Additionally, we created stealth assessment of causal reasoning in the *World of Goo* (Shute & Kim, 2011) and systems thinking in *Taiga Park* (Shute, Masduki, & Donmez, 2010). From these experiences, we have derived some general steps related to the design and development of stealth assessment, shown in the 9-step approach listed as follows. In the following section, we illustrate how we implemented these steps using two recent research projects.

1. Develop competency model (CM) of targeted knowledge, skills, or other attributes based on full literature and expert reviews
2. Determine which game (or learning environment) the stealth assessment will be embedded into
3. Delineate a full list of relevant gameplay actions/indicators that serve as evidence to inform CM and its facets
4. Create new tasks in the game, if necessary (Task model, TM)
5. Create Q-matrix to link actions/indicators to relevant facets of target competencies
6. Determine how to score indicators using classification into discrete categories (e.g., yes/no, very good/good/ok/poor relative to quality of the actions). This becomes the "scoring rules" part of the evidence model (EM)
7. Establish statistical relationships between each indicator and associated levels of CM variables (EM)
8. Pilot test Bayesian Networks (BNs) and modify parameters
9. Validate the stealth assessment with external measures

4.2.3 Adaptation

The next logical step—which is currently under development—involves using the current information about a player's competency states to provide adaptive learning support (e.g., targeted formative feedback, progressively harder levels relative to the player's abilities, and so on). The adaptive difficulty features in a video game may potentially increase motivation and enhance learning by providing the right level of challenge (i.e., tasks that are neither too easy nor too difficult). Such optimal levels of challenge ensure that the learner is kept in the zone of proximal development (ZPD). Within ZPD, learning activities are just beyond the learner's ability but can be achieved with guidance (Vygotsky, 1978). The guidance is sometimes referred to as instructional scaffolding. Some examples of such scaffolding include targeted formative feedback and hints to help learners proceed in the task. Studies show that scaffolded learning activities lead to better learning outcomes compared with activities without scaffolds (e.g., Chang, Sung, & Chen, 2001; Murphy & Messer, 2000). In addition, when tasks are too complicated for a learner, he or she may encounter cognitive overload that exceeds the capacity of their working memory and thus undermines learning. On the other hand, if the tasks are too easy, the learner may feel bored and disengaged, which also negatively affects learning. Therefore, it is important and beneficial to adjust the difficulty of tasks to the competencies of the individual and provide appropriate learning scaffolds.

There are two main approaches to produce adapted content in video games—offline and online adaptivity (Lopes & Bidarra, 2011). For offline adaptivity, content is adjusted after gathering sufficient information about the learner before he or she starts playing the game. For online adaptivity (or dynamic adaptivity; see van Oostendorp, van der Spek, & Linssen, 2014), the content is adjusted based on learner's performance, in real time. We recommend the second approach because the assessment of the learner's competency will be more accurate when he or she is actually performing the task.

Some common ways to gather information about the learner during gameplay include the use of infrared camera or emotion detection software, and stealth assessment. One issue with infrared camera or emotion detection software is that different people may experience different levels of stress when they are under pressure. Thus, it is difficult to choose the right task based on the stress level. Alternatively, stealth assessment gathers data unobtrusively based on performance in the game and is free from such bias.

To determine the sequence of tasks in video games, researchers have attempted to set an agreed-upon threshold value (e.g., level up after three consecutive successes; see Sampayo-Vargas, Cope, He, & Byrne, 2013). Some have calculated the expected weight of evidence to pick tasks that will maximize the information about a player (Shute et al., 2008). Due to the relatively high cost of developing adaptive educational games, few researchers have attempted to investigate the effects of adaptive video games on learning. However, existing evidence shows that such methods are promising. For example, van Oostendorp et al. (2014) compared the effects of an adaptive version of a game focusing on triage training against a version without adaptation. They reported that those who played the adaptive version of the game learned better than those in the control group.

4.3 Examples of Stealth Assessment

4.3.1 "Use Your Brainz" (UYB)

4.3.1.1 Competency Model Development and Game Selection (Steps 1 and 2)

In the UYB project, we developed a stealth assessment of problem-solving skills and embedded it within the modified version of the commercial game *Plants* vs. *Zombies 2* (the education version is called "Use your Brainz"). The project was a joint effort between our research team and GlassLab. PvZ 2 is a tower defense type of game. The goal is to protect the home base from the invasion of zombies by planting various defensive and offensive plants in the limited soil in front of the home base. We selected 43 game levels arranged by difficulty. Figure 4.1 shows an example of one of the levels in the game.

We chose the game PvZ 2 for two main reasons. First, the game provides a meaningful and engaging context where players are expected to acquire knowledge about the rules of the game and apply different resources in the game to solve intriguing problems. Second, GlassLab had access to the source code from EA—the publisher of PvZ 2—which enabled us to customize the log files.

Fig. 4.1 Screen capture of UYB gameplay on Level 9, World 1 (Ancient Egypt)

After we determined that we would like to model problem-solving skills, we reviewed the literature on how other researchers have conceptualized and operationalized problem solving. In addition to our extensive review of the literature on problem-solving skills, we also reviewed the Common Core State Standards (CCSS) related to problem solving. We came up with a four-facet competency model (CM), which included: (a) understanding givens and constraints, (b) planning a solution pathway, (c) using tools effectively/efficiently when implementing solutions, and (d) monitoring and evaluating progress.

4.3.1.2 Identifying Gameplay Indicators (Steps 3 and 4)

Our next task entailed identifying specific in-game behaviors that would serve as valid evidence and thus inform the status of the four-facet competency model. After playing the game repeatedly and watching expert solutions on YouTube, we delineated 32 observable indicators that were associated with the four facets. For example, sunflowers produce sun power, which is the sole source of power that players may use to grow plants. At the beginning of a level, typically there are no or very few sunflowers on the battlefield. To supply power to grow plants, players must plant sunflowers at the beginning of each level before zombies start to appear in waves. After brainstorming with the PvZ 2 experts on our research team, we decided that the scoring rule for this particular indicator was: "*If a player plants more than three sunflowers before the second wave of zombies arrives, the student understands the time and resource constraints.*" Table 4.1 displays a sample of indicators for each of the four problem-solving facets. Overall, we included 7 indicators for "analyzing givens and constraints," 7 for "planning a solution pathway," 14 for "using tools effectively and efficiently," and 4 for "monitoring and evaluating progress." The list of indicators forms our task model and the scoring rules form a part of the evidence model.

Table 4.1 Examples of indicators for each problem-solving facet

Facet	Example indicators
Analyzing givens and constraints	• Plants >3 Sunflowers before the second wave of zombies arrives
	• Selects plants off the conveyor belt before it becomes full
Planning a solution pathway	• Places sun producers in the back/left, offensive plants in the middle, and defensive plants up front/right
	• Plants Twin Sunflowers or uses plant food on (Twin) Sunflowers in levels that require the production of X amount of sun
Using tools effectively and efficiently	• Uses plant food when there are >5 zombies in the yard or zombies are getting close to the house (within two squares)
	• Damages >3 zombies when firing a Coconut Cannon
Monitoring and evaluating progress	• Shovels Sunflowers in the back and replaces them with offensive plants when the ratio of zombies to plants exceeds 2:1

4.3.1.3 Q-Matrix Development and Scoring Rules (Steps 5 and 6)

We created a Q-matrix (Almond, 2010; Tatsuoka, 1990) laying out all of the indicators in rows and the four facets in the columns. We added a "1" in the crossed cell if the indicator was relevant to the facet and "0" if the facet did not apply to the indicator. We then went through each indicator and discussed how we could classify each indicator into discrete scoring categories such as "yes/no" or "very good/good/ok/poor." The overall scoring rules were based on a tally of relevant instances of observables. Using the aforementioned sunflower indicator, if a player successfully planted more than three sunflowers before the second wave of zombies arrived on the scene, the log file would automatically record the action and categorize it as a "yes" status of the indicator.

For another example, consider the facet "using tools effectively and efficiently." In Table 4.1, an example indicator is "uses plant food when there are >5 zombies in the yard or zombies are getting close to the house (within two squares)." Plant food in the game is a rare resource. Using one dose of plant food on any plant will substantially boost the effect of the plant—whether offensive or defensive—for a short period of time. This indicator would be scored if the player used plant food as a boost (a) when there were more than five zombies on the battlefield, or (b) when zombies were within two squares in front of the house (where the overarching goal of each level is to protect the house from zombies). Since a single instance of this "using plant food" action may be performed by chance, the completion status of the indicator was categorized into four levels. That is, the game engine checks on the ratio of the indicator, which is "the number of times that plant food was used when >5 zombies in the yard or within two squares in front of the house, divided by the total number of times that plant food was used in the level." Then the game engine maps the value of the ratio onto one of the four states of the indicator where in this case, higher means better. If the value is within [0, 0.25], it corresponds to the status of "poor" performance on the indicator; if the value falls within [0.26, 0.5], it corresponds to the "ok" status; if the value falls within [0.51, 0.75], it corresponds to the "good" status, and if the ratio falls within [0.76, 1], it is categorized as "very good."

4.3.1.4 Establishing Statistical Relationships Between Indicators and CM Variables (Step 7)

Once we categorized all indicators into various states, we needed to establish statistical relationships between each indicator and the associated levels of the CM variables. We used Bayesian networks (BNs) to accumulate incoming data from gameplay and update beliefs in the CM. The relationship between each indicator and its associated CM variable was expressed within conditional probability tables stored in each Bayes net. We created a total of 43 Bayes nets for this project, one for each level. We used separate BNs because many indicators do not apply in every level and computations would be more efficient for simpler networks. The statistical relationships carried in the Bayes nets and the scoring rules described in the last section formed the evidence model.

Table 4.2 Conditional probability table for indicator #8 "plant >3 sunflowers before the second wave of zombies" in Level 9

Analyzing givens and constraints	Yes	No
High	.82	.18
Medium	.73	.27
Low	.63	.37

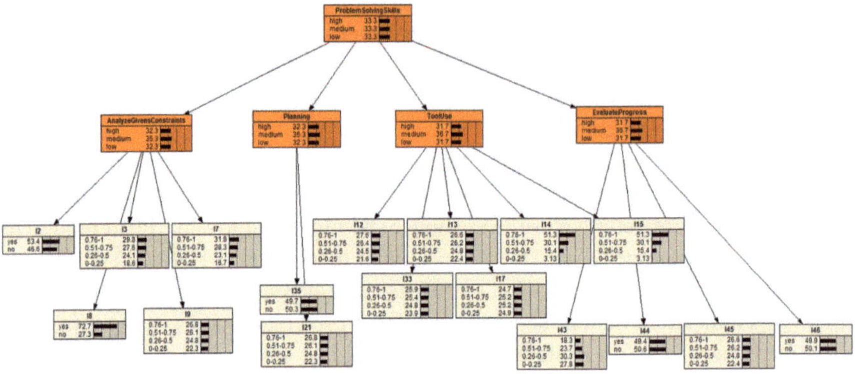

Fig. 4.2 Bayes network of level 9 in UYB, prior probabilities

Table 4.2 shows the conditional probability table we created for indicator #8, "Plants >3 Sunflowers before the second wave of zombies arrives" (associated with the facet "analyzing givens and constraints") in Level 9. Because the game is linear (i.e., you need to solve the current level before moving to the next level), by the time a player gets to Level 9, she has had experience playing previous levels, thus should be quite familiar with the constraint of planting sunflowers at this point. Consequently, this indicator should be relatively easy to accomplish (i.e., the probabilities to fail the indicator were low despite one's ability to analyze givens and constraints). Even those who are low on the facet still have a probability of .63 of accomplishing this indicator. When evidence about a student's observed results on indicator #8 arrives from the log file, the estimates on his ability to analyze givens and constraints will be updated based on Bayes theorem. We configured the distributions of conditional probabilities for each row in Table 4.2 based on Samejima's graded response model, which includes the item response theory parameters of discrimination and difficulty (see Almond, 2010; Almond et al., 2001; Almond, Mislevy, Steinberg, Williamson, & Yan, 2015). In this case, the difficulty was set at −2 (very easy) and the discrimination value was 0.3 (i.e., may not separate students with high versus low abilities well).

As a player interacts with the game, incoming evidence about the player's status on certain indicators updates the estimates about relevant facets. The evidence then propagates through the whole network and thus estimates related to student problem-solving skills are updated. The Bayes nets keep accumulating data from the indicators and updating probability distributions of nodes in the network. For example, Fig. 4.2 displays a full Bayes net of Level 9 prior probabilities (see Fig. 4.1 for an illustration of the level). Shaded nodes toward the top are the competency

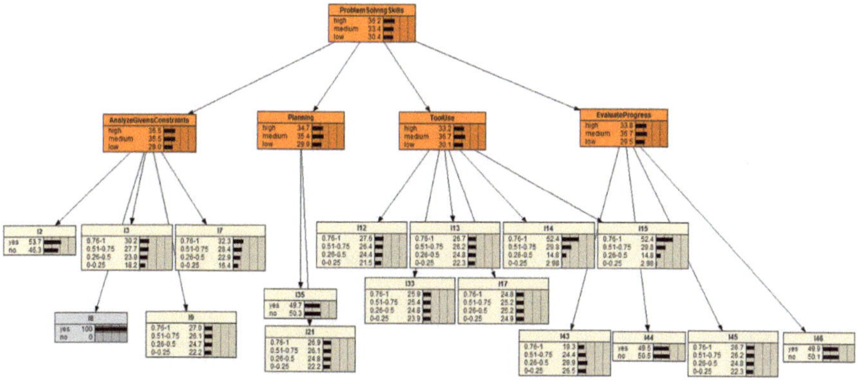

Fig. 4.3 Evidence of the completion of indicator #8

variables, while the beige nodes toward the bottom represent all relevant indicators. We used the program Netica (by Norsys Software Corporation) to construct and compile the network.

For instance, if a player successfully completed indicator #8 in Level 9 (i.e., planting sufficient sunflowers prior to a wave of incoming zombies), the log file records the action, informs the network of the new evidence, and the data are propagated throughout the network (see Fig. 4.3). As shown, the updated probability distribution of the player's level of "analyzing givens and constraints" is: Pr (analyzing givens and constraints | high) = .365, Pr (analyzing givens and constraints | med) = .355, Pr (analyzing givens and constraints | low) = .280. The estimates for the player's overall problem-solving skill are Pr (problem solving | high) = .362, Pr (problem solving | med) = .334, Pr (problem solving | low) = .304. Because there is no clear modal state for the problem-solving skills node (i.e., the difference between high and medium states is just .028), this suggests that more data are needed.

Alternatively, suppose the player fails to accomplish the indicator by the second wave of zombies. In this case, the log file would record the failure, inform the BN of the evidence, and update with new probability distributions for each node (Fig. 4.4). The current probability distribution of the player's level of "analyzing givens and constraints" is Pr (analyzing givens and constraints | high) = .213, Pr (analyzing givens and constraints | med) = .349, Pr (analyzing givens and constraints | low) = .438. The estimates for the player's overall problem solving skill are Pr (problem solving | high) = .258, Pr (problem solving | med) = .331, Pr (problem solving | low) = .411. This shows that the student is likely to be low in relation to problem-solving skills.

4.3.1.5 Pilot Testing Bayes Nets (Step 8)

Our game experts and psychometricians produced the initial prior probabilities of each node in each network collaboratively. We hypothesized that students would have an equal likelihood of being "high," "medium," or "low" on problem solving

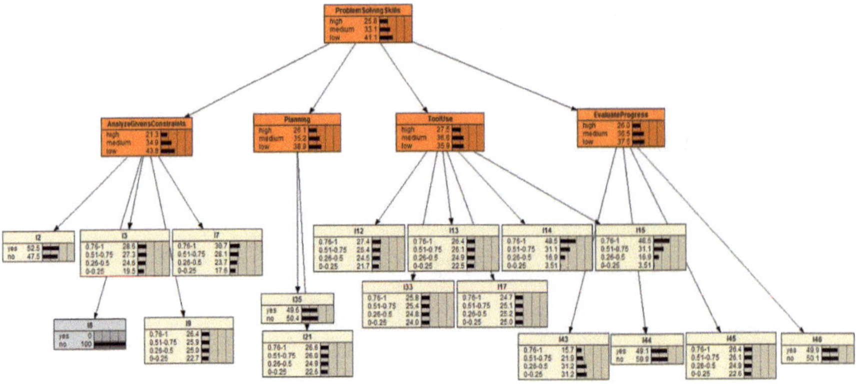

Fig. 4.4 Evidence of failure to complete indicator #8

and the probability of being "high," "medium," or "low" for each facet would be normally distributed. As more evidence enters the network, the estimates become more accurate and tend to reflect each player's true status on the competency. After developing the BNs and integrating them into the game code, we were able to acquire real-time estimates of players' competency levels across the main node (problem-solving skill) and its constituent facets. We acknowledge that any initial probabilities may be subject to bias or inaccurate judgment. Therefore, we ran a pilot test and used the ensuing pilot data to adjust parameters of the Bayes nets accordingly.

4.3.1.6 Validating Stealth Assessment (Step 9)

The final step in our list of stealth assessment processes is the validation of the stealth assessment against external measures. For the UYB project, we employed two external measures: *Raven's Progressive Matrices* (Raven, 1941, 2000) and *MicroDYN* (Wustenberg, Greiff, & Funke, 2012). Raven's is a test that examines subjects' ability to reason based on given information. MicroDYN presents to subjects a simulation system where subjects are expected to acquire and apply information. For a thorough overview on MicroDYN, see Schweizer, Wüstenberg, and Greiff (2013) and Wustenberg, Greiff, and Funke (2012).

We recruited 55 7th grade students from a middle school in suburban Illinois. Students played UYB for 3 h (1 h per day across three consecutive days) and completed the external measures on the fourth day. Among the 55 participants, one student's gameplay data was missing, five students did not take the Raven's test, and two students did not complete the MicroDYN test. After we removed the missing data, we had complete data from 47 students (20 male, 27 female).

Results show that our game-based stealth assessment of problem-solving skills is significantly correlated with both Raven's ($r = .40$, $p < .01$) and MicroDYN ($r = .41$, $p < .01$), which established the construct validity of our stealth assessment. We are also refining our Bayes nets based on data collected. These test results need to be verified with an even larger sample.

This example demonstrates step by step how we modeled problem-solving skills and created and implemented stealth assessment of the skill in the context of a modified commercial game. Specifically, we created our competency model of problem-solving skills based on the literature, identified relevant indicators from gameplay that could provide evidence of players' levels on the competency model variables, crafted scoring rules of each indicator, and connected the indicators statistically with competency model variables. We then modified the Bayes networks by collecting and analyzing data collected from a pilot study. Then, we selected well-established external measures and validated the stealth assessment in a validation study. Reasonable next steps would entail developing tools to help educators gain access to the results of the assessment easily (e.g., via a dashboard displaying and explaining important results). With that information, educators could effectively and efficiently support the growth of problem-solving skill, at the facet level.

4.3.2 "Earthquake Rebuild" (E-Rebuild)

As discussed in the preceding example with UYB, the stealth assessment was designed and implemented as a post-hoc practice because the game had already been designed. In a current design-based project (called Earthquake Rebuild), we have been designing evidenced-centered stealth assessment during the entire course of game design. Earthquake Rebuild (E-Rebuild) acts as both a testbed and sandbox for generating, testing, and refining the focus design conjectures on game-design-associated, stealth assessment and support of learning.

Developed using Unity 3D, the overall goal of E-Rebuild is to rebuild an earthquake-damaged space to fulfill diverse design parameters and needs. The intermediate game goal involves completing the design quest(s) in each game episode to gain new tools, construction materials, and credits. A learner in E-Rebuild performs two modes of play: (a) third-person construction mode, and (b) first-person adventure mode. In the third-person construction mode, a learner performs construct site survey and measurement and maneuver (e.g., cut/scale, rotate, and stack up) construction items to build the targeted structure. In the adventure mode, a learner navigates the virtual world, collects or trade construction items, and assigns space (to residents, for example).

The process of interweaving game and assessment design in E-Rebuild included four core design sectors: (1) developing competency models and selecting game mechanics that necessitate the performance of the focus competency, (2) designing game task templates and contextual scenarios along with the Q-matrix, (3) designing the game log file based on the Q-matrix, and (4) designing the in-game support as both live input for data-driven assessment and adaptive feedback. These design sectors are interacting and interdependent with each other.

4.3.2.1 Competency Model and Game Mechanics Development

In E-Rebuild, an interdisciplinary team of math educator, mathematician, and assessment experts codeveloped a competency model for each focal math topic. These competency models are aligned with the Common Core State Standards (CCSS) for mathematical practice in grades 6–8. The game design team then designed and selected game mechanics that would best serve the competency models. Specifically, game actions were the core constituent of game mechanics and the basic behavioral unit to be tracked during gameplay. Consequently, game actions became the driving element, defining the degree of learning integration and assessment in the game. The team focused on designing game actions or indicators that would *necessitate*, *not just allow*, the performance of focus knowledge and skills (e.g., ratio and proportional reasoning). By experimenting with all proposed architectural design actions via iterative expert review and user testing at the initial paper prototyping stage, the design team decided on the game actions that best operationalized the practice of math knowledge, which include (material) trading, building, and (resource) allocation. Furthermore, comparative analyses with different versions of the game prototype in a one-year case study indicated that an intermediary yet noninterruptive user input (e.g., entering a specific number), in comparison with an intuitive user input (e.g., clicking or dragging a button or meter to adjust a numerical value), effectively necessitates the practice of the targeted mathematical knowledge. For example, the trading interface (see Fig. 4.5) requires the player to enter the quantity of a building item to be ordered, calculate the total amount/cost (based on the unit rate), and enter the numerical value. Similarly, the scaling tool prompts the player to specify the numerical value for the scaling factor to scale down a 3D construction item along the chosen local axis of the item (*x*, *y*, *z*, or all).

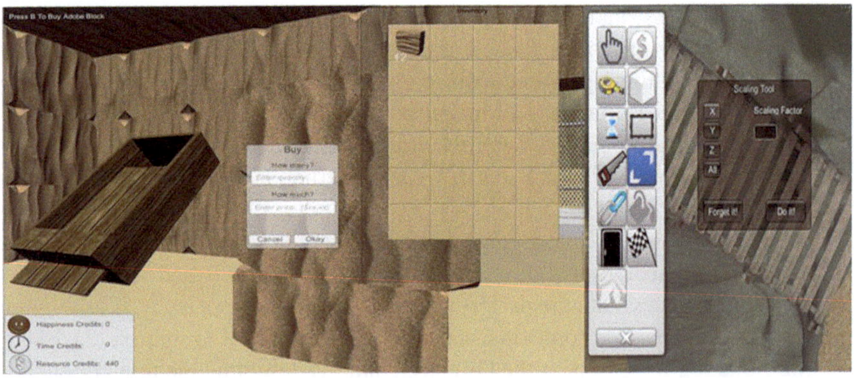

Fig. 4.5 Intermediary user input example—the trading interface and the scaling tool for the building action

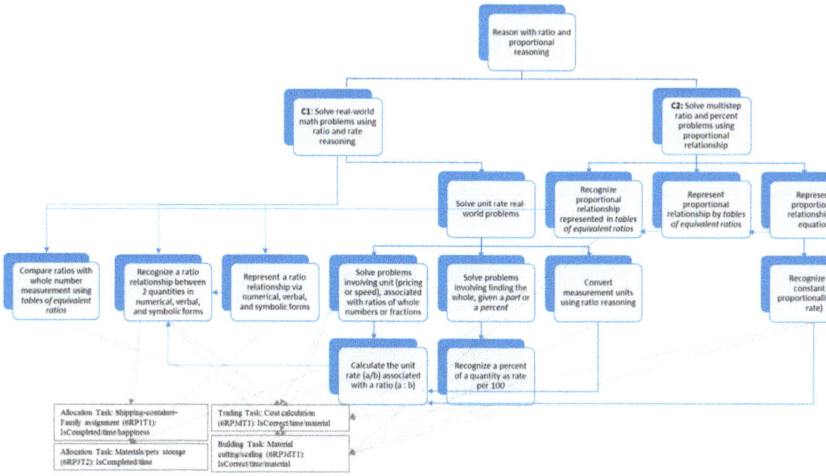

Fig. 4.6 A design document depicting a competency model along with the design of game task templates. *Note*: The four *black boxes* at the bottom represent examples of game tasks designed to extract the subcompetencies, which are depicted in the *blue boxes* in a hierarchical structure. *Solid lines* indicate the relationships among competencies and subcompetencies to be captured/assessed, and *dotted lines* link the gaming tasks and the competencies to be assessed.

4.3.2.2 Designing Task Templates to Substantiate the Competency Model and Q-Matrix

In E-Rebuild, the game task development was confined by the math competency models. Specifically, the competency model has driven the development of a cluster of game task templates and the selection of the tasks' parameters and content scope (as depicted in Fig. 4.6). For instance, an exemplary allocation task (e.g., assigning families into a multiroom shelter structure, with the ratio of an adult's living space need to a child's need being 2 to 1) was designed to extract math performance of subcompetencies (e.g., C1) of "ratio and proportional reasoning." The Q-matrix development (Fig. 4.7) then helped the design team gauge and track which facets of the math competency a specific gameplay action inform, and whether each facet of a math competency is practiced/assessed by different clusters of tasks. Accordingly, existing task templates could be refined or removed, and new task templates might be developed.

The Q-matrix also helped the team to gauge the discrimination and difficulty qualities of different tasks and hence assisted the selection and sequencing of tasks within/across game episodes. Finally, a variety of architecture-themed scenarios (e.g., building shelters with shipping containers or building a structure to meet the needs of multiple families) would contextualize different clusters of game tasks and inform the development of the task narrative. These aforementioned design processes occurred concurrently and helped to make the game-task design and the evidence model development a coherent process.

Task Name	ObsName	Reason with ratio and proportional reasoning								
		Compare ratios with whole number measurement using tables of equivalent ratios	Recognize a ratio relationship between 2 quantities in numerical form	Recognize a ratio relationship between 2 quantities in verbal form	Recognize a ratio relationship between 2 quantities in symbolic form	Represent a ratio relationship via numerical form	Represent a ratio relationship via verbal form	Represent a ratio relationship via symbolic form	Calculate the unit rate (a/b) associated with a ratio (a : b)	Recognize a percent of a quantity as rate per 100
Allocation Task	timeToCompletion	0	1	1	1	1	0	1	1	0
	Material Credit	0	0	0	0	0	0	0	1	0
	scratchpad editing(math related)	0	0	0	0	1	0	0	1	0
	assignment operation	0	0	0	1	0	0	1	1	0
Trading Task	# of trades	1	1	1	0	1	0	0	1	1
	scratchpad editing(math related)	0	0	0	0	1	0	0	1	0
	percentage lost in trade avg	1	1	1	0	1	0	0	1	1
	cut (for resourcing)	0	0	0	0	0	0	0	0	0
	scale (for resourcing)	0	0	0	0	0	0	0	0	0
Building Task	structure size	0	0	1	0	0	0	1	0	1
	structure location	0	0	0	0	0	0	0	0	0
	structure direction	0	0	0	0	0	0	0	0	0
	# copy/paste failed	0	0	0	0	0	0	0	0	0
	scratchpad editing(math related)	0	0	0	0	1	0	0	1	0
	ruler record	0	0	1	0	0	0	1	0	0
Game Task	timeToCompletion	1	1	1	1	1	0	1	1	1
	Material Credit	1	1	1	1	1	0	1	1	1
	Happiness Credit	0	1	1	1	1	0	1	1	1

Fig. 4.7 Part of the Q-matrix for E-Rebuild. *Note*: Facets of the focus competency are listed in columns and the indicators are listed in rows.

4.3.2.3 Designing Game Log File Along with Q-Matrix for Bayesian Network Construction

During the course of E-Rebuild design, we designed, tested, and refined the game log file along with the Q-matrix so that the game objects, salient object features, play actions, and action-performing statuses tracked in the game log will assist the generation and update of conditional probability tables (CPTs) for all indicators in the Bayes net being constructed. In E-Rebuild, the creation of CPTs for indicators and hence the Bayesian Network construction were initially driven by the logged gameplay data of 42 middle school students and 6 game/content experts in a pilot study. The CPTs and the preliminary networks generated were then reviewed and refined by the content/assessment experts and game designers. Game logs and indicators were also refined based on the pilot-testing results. For the next phase, the refined CPTs and Bayesian networks will be further tested and updated by the gameplay data to be collected from a larger group of target users, and then validated by external math knowledge tests in a future evaluation study.

4.3.2.4 *In-Game Support as Both Input and Output of Data-Driven Learning Assessment*

In E-Rebuild, we have designed in-game cognitive support (scaffolding) as an expandable/collapsible help panel and a scratch pad. The scratch pad includes an internal calculator and enables/records participants' typing of numerical calculation steps. The help panel (Fig. 4.8) contains interactive probes to facilitate active math problem representation rather than passively presenting the information. When

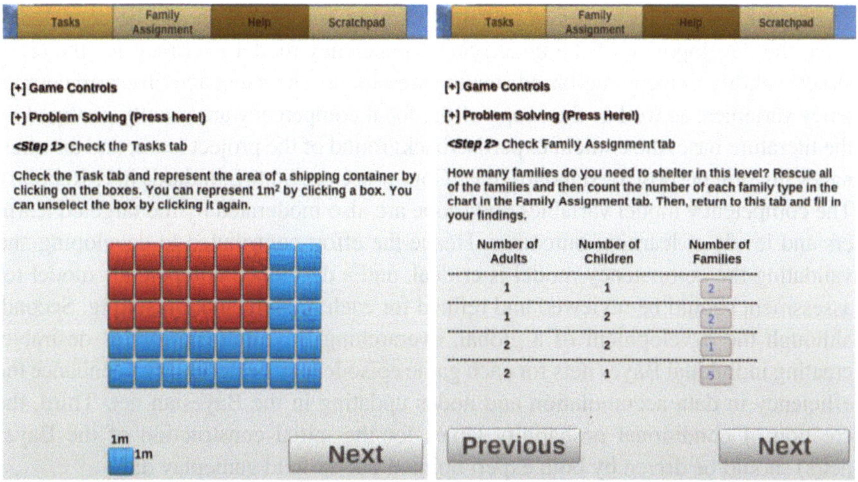

Fig. 4.8 Interactive learning probes

interacting with those probes, a player has to enter numbers or maneuver dynamic icons, with all interactions logged. The two support features thus work as another dynamic data source for game-based stealth assessment. In addition, we are still designing the dynamic-help mechanism that will use the values extracted from the logged gameplay performance variables (e.g., timeToCompletion, materialCredit, assignmentScore, usedScratchpad, helpInput) to inform the content and presentation of task-specific learner feedback in the Help panel. Based on the dynamically updated game task performance of the player, the game-based assessment mechanism will inform on task-relevant math competency (e.g., below 50 % in a specific competency). Accordingly, the help menu will be displayed automatically and a math-competency-related subsection of the problem-solving probes will be expanded. The interactive probes may be presented in iconic (pictorial) and/or symbolic (numerical formula) formats, pending on the player's choice.

4.4 Discussion and Implications

In this chapter, we have introduced the core steps of game-based stealth assessment of learning and illustrated the implementation of these steps via two cases. The first case focuses on developing an assessment mechanism for an existing game and the assessment of an important domain-general skill (i.e., problem solving). The second case highlights the integration of learning task and assessment design throughout the game development process and the assessment of domain-specific (mathematical) practice and learning. Both cases illustrate the applicability of data-driven, performance-based assessment in an interactive learning setting, for either formal or informal learning.

Several design challenges of in-game learning assessment should be considered. First, the development of the underlying competency model is critical for the (construct) validity of the game-based stealth assessment. The latent and observed competency variables, as well as the scope of the focal competency are usually confined by the literature base, the content expertise/background of the project team, and an external evaluation purpose or standard (e.g., Common Core State Standards in E-Rebuild). The competency model variables and scope are also moderated by the targeted learners and levels of learning outcomes. Hence the effort contributed to developing and validating the competency model is critical, and a developed competency model for assessment should be reviewed and refined for each implementation setting. Second, although the development of a global, overarching Bayesian network is desirable, creating individual Bayes nets for each game episode may be necessary to enhance the efficiency in data accumulation and nodes updating in the Bayesian net. Third, the creation of conditional probability tables for the initial construction of the Bayes net(s) should be driven by both expert opinion and in-field gameplay data.

In the first game (Use Your Brain), expert opinions drove the initial CPT development, which were then enhanced by in-field data validation. In E-Rebuild, CPTs were generated (learned) from the in-field data and then reviewed/refined by experts. Future research can experiment with the two methods in CPT generation and further investigate the potential differences in the two methods on learning and validating the Bayesian network. Finally, in both projects we are presently developing and testing various adaptive learning support mechanisms. The dynamically updated learning assessment in E-Rebuild will be used to drive the timing (e.g., at the end of a game action, a task, or a game level), topic (e.g., on a task-specific math concept or a calculation procedure), and the presentation format (e.g., iconic or symbolic, informative hint or interactive probe) of the learning scaffolds for game-based learning. A critical design consideration for assessment-based, dynamic learner support is the timing and extent of live data accumulation for adaptive support presentation. In E-Rebuild, we have used game level and game episode (i.e., an episode includes multiple game levels) as two hierarchical units for data accumulation and learning support presentation. Specifically, performance data will be fed into the Bayesian network at the end of each game level and each game episode. Correspondingly, the learner profile will be updated at these points, and then the relevant learner supports (e.g., probes and feedback) can be presented as both cut-screen in between game levels/episodes, and updated content in the Help panel.

References

Almond, R. G. (2010). Using evidence centered design to think about assessments. In V. J. Shute & B. J. Becker (Eds.), *Innovative assessment for the 21st century: Supporting educational needs* (pp. 75–100). New York: Springer.

Almond, R. G., DiBello, L., Jenkins, F., Mislevy, R. J., Senturk, D., Steinberg, L. S., et al. (2001). Models for conditional probability tables in educational assessment. In T. Jaakkola & T. Richardson (Eds.), *Artificial intelligence and statistics 2001* (pp. 137–143). San Francisco, CA: Morgan Kaufmann.

Almond, R. G., Mislevy, R. J., Steinberg, L. S., Williamson, D. M., & Yan, D. (2015). *Bayesian networks in educational assessment*. New York: Springer.

Barab, S. A., Gresalfi, M., & Ingram-Goble, A. (2010). Transformational play using games to position person, content, and context. *Educational Researcher, 39*(7), 525–536.

Barab, S. A., Thomas, M., Dodge, T., Carteaux, R., & Tuzun, H. (2005). Making learning fun: Quest Atlantis, a game without guns. *Educational Technology Research and Development, 53*(1), 86–108.

Bransford, J., Brown, A. L., & Cocking, R. R. (2000). *How people learn: Brain, mind, experience, and school* (expanded ed.). Washington: National Academies Press.

Chang, K. E., Sung, Y. T., & Chen, S. F. (2001). Learning through computer-based concept mapping with scaffolding aid. *Journal of Computer Assisted Learning, 17*, 21–33.

Clark, D. B., Tanner-Smith, E. E., & Killingsworth, S. (2014). *Digital games, design, and learning: A systematic review and meta-analysis*. Menlo Park, CA: SRI International.

Cowley, B., Charles, D., Black, M., & Hickey, R. (2008). Toward an understanding of flow in video games. *Computers in Entertainment, 6*(2), 1–27.

Csikszentmihalyi, M. (1990). *Flow: The psychology of optimal experience*. New York: Harper & Row.

Csikszentmihalyi, M. (1997). *Finding flow*. New York: Basic.

DiCerbo, K. E., & Behrens, J. T. (2012). Implications of the digital ocean on current and future assessment. In R. Lissitz & H. Jiao (Eds.), *Computers and their impact on state assessment: Recent history and predictions for the future* (pp. 273–306). Charlotte, NC: Information Age Publishing.

Embretson, S. E. (1998). A cognitive design system approach to generating valid tests: Application to abstract reasoning. *Psychological Methods, 3*(3), 300–396. doi:10.1037/1082-989X.3.3.380.

Entertainment Software Association. (2015). *2015 Essential facts about the computer and video game industry*. Retrieved from http://www.theesa.com/wp-content/uploads/2015/04/ESA-Essential-Facts-2015.pdf

Flynn, L. (2008). In praise of performance-based assessments. *Science and Children, 45*(8), 32–35.

Fullerton, T. (2014). *Game design workshop, 3rd edition: A playcentric approach to creating innovative games*. Boca Raton, FL: AK Peters/CRC Press.

Gee, J. P. (2003). *What video games have to teach us about learning and literacy*. New York: Palgrave Macmillan.

Kim, Y. J., & Shute, V. J. (2015). Opportunities and challenges in assessing and supporting creativity in video games. In J. Kaufmann & G. Green (Eds.), *Research frontiers in creativity*. San Diego, CA: Academic.

Lopes, R., & Bidarra, R. (2011). Adaptivity challenges in games and simulations: A survey. *IEEE Transactions on Computational Intelligence and AI in Games, 3*(2), 85–99.

Madaus, G. F., & O'Dwyer, L. M. (1999). A short history of performance assessment: Lessons learned. *Phi Delta Kappan, 80*(9), 688–695.

Mislevy, R. J., Steinberg, L. S., & Almond, R. G. (2003). On the structure of educational assessment. *Measurement: Interdisciplinary Research and Perspective, 1*(1), 3–62.

Murphy, N., & Messer, D. (2000). Differential benefits from scaffolding and children working alone. *Educational Psychology, 20*(1), 17–31.

Partnership for the 21st Century. (2015). Retrieved from http://www.p21.org/storage/documents/P21_framework_0515.pdf

Raven, J. C. (1941). Standardization of progressive matrices, 1938. *British Journal of Medical Psychology, 19*(1), 137–150.

Raven, J. (2000). The Raven's progressive matrices: Change and stability over culture and time. *Cognitive Psychology, 41*, 1–48.

Sampayo-Vargas, S., Cope, C. J., He, Z., & Byrne, G. J. (2013). The effectiveness of adaptive difficulty adjustments on students' motivation and learning in an educational computer game. *Computers & Education, 69*, 452–462.

Schweizer, F., Wüstenberg, S., & Greiff, S. (2013). Validity of the MicroDYN approach: Complex problem solving predicts school grades beyond working memory capacity. *Learning and Individual Differences, 24*, 42–52.

Shute, V. J. (2008). Focus on formative feedback. *Review of Educational Research, 78*(1), 153–189.
Shute, V. J. (2011). Stealth assessment in computer-based games to support learning. In S. Tobias & J. D. Fletcher (Eds.), *Computer games and instruction* (pp. 503–524). Charlotte, NC: Information Age Publishers.
Shute, V. J., Hansen, E. G., & Almond, R. G. (2008). You can't fatten a hog by weighing it—Or can you? Evaluating an assessment for learning system called ACED. *International Journal of Artificial Intelligence and Education, 18*(4), 289–316.
Shute, V. J., & Kim, Y. J. (2011). Does playing the World of Goo facilitate learning? In D. Y. Dai (Ed.), *Design research on learning and thinking in educational settings: Enhancing intellectual growth and functioning* (pp. 359–387). New York, NY: Routledge Books.
Shute, V. J., Leighton, J. P., Jang, E. E., & Chu, M.-W. (2016). Advances in the science of assessment. *Educational Assessment., 21*(1), 1–27.
Shute, V. J., Masduki, I., & Donmez, O. (2010). Conceptual framework for modeling, assessing, and supporting competencies within game environments. *Technology, Instruction, Cognition and Learning, 8*(2), 137–161.
Shute, V. J., & Ventura, M. (2013). *Measuring and supporting learning in games: Stealth assessment*. Cambridge, MA: The MIT Press.
Shute, V. J., Ventura, M., & Ke, F. (2015). The power of play: The effects of Portal 2 and Lumosity on cognitive and noncognitive skills. *Computers & Education, 80*, 58–67.
Shute, V. J., Ventura, M., & Kim, Y. J. (2013). Assessment and learning of qualitative physics in Newton's Playground. *The Journal of Educational Research, 106*, 423–430.
Shute, V. J., & Wang, L. (2016). Assessing and supporting hard-to-measure constructs. In A. A. Rupp, & J. P. Leighton (Eds.), *The handbook of cognition and assessment: Frameworks, methodologies, and application* (pp. 535–562). Hoboken, NJ: John Wiley & Sons, Inc.
Sitzmann, T. (2011). A meta-analysis of self-regulated learning in work-related training and educational attainment: What we know and where we need to go. *Psychological Bulletin, 137*, 421–442.
Squire, K. (2006). From content to context: Videogames as designed experience. *Educational Researcher, 35*(8), 19–29.
Stecher, B. (2010). *Performance assessment in an era of standard-based educational accountability*. Stanford, CA: Stanford University, Stanford Center for Opportunity Policy in Education.
Tatsuoka, K. (1990). Toward an integration of item-response theory and cognitive error diagnosis. In N. Frederiksen, R. Glaser, A. Lesgold, & M. Shafto (Eds.), *Diagnostic monitoring of skill and knowledge acquisition* (pp. 453–488). Hillsdale, NJ: Erlbaum.
van Oostendorp, H., van der Spek, E. D., & Linssen, J. (2014). Adapting the complexity level of a serious game to the proficiency of players. *EAI Endorsed Transactions on Serious Games, 1*(2), 8–15.
Ventura, M., Shute, V. J., & Small, M. (2014). Assessing persistence in educational games. In R. Sottilare, A. Graesser, X. Hu, & B. Goldberg (Eds.), *Design recommendations for adaptive intelligent tutoring systems* (Learner modeling, Vol. 2, pp. 93–101). Orlando, FL: U.S. Army Research Laboratory.
Ventura, M., Shute, V. J., & Zhao, W. (2012). The relationship between video game use and a performance-based measure of persistence. *Computers and Education, 60*, 52–58.
Vygotsky, L. S. (1978). *Mind in society: The development of higher mental processes*. Cambridge, MA: Harvard University Press.
Vygotsky, L. S. (1987). *The collected works of L. S. Vygotsky*. New York: Plenum.
Wouters, P. J. M., van Nimwegen, C., van Oostendorp, H., & van der Spek, E. D. (2013). A meta-analysis of the cognitive and motivational effects of serious games. *Journal of Educational Psychology, 105*, 249–265.
Wustenberg, S., Greiff, S., & Funke, J. (2012). Complex problem solving—More than reasoning? *Intelligence, 40*, 1–14.

Chapter 5
Fidelity and Multimodal Interactions

Bill Kapralos, Fuad Moussa, Karen Collins, and Adam Dubrowski

Abstract Often, designers and developers of serious games (and virtual simulations in general) strive for high fidelity (realism). However, real-time high-fidelity rendering of complex environments across sensory modalities such as vision, audition, and haptic (sense of touch) is still beyond our computational reach. Previous work has demonstrated that multimodal effects can be considerable, to the extent that a large amount of detail of one sense may be ignored or enhanced by the presence of other sensory inputs. Taking advantage of such multimodal effects, *perceptual-based rendering*—whereby the rendering parameters are adjusted based on the perceptual system—can be employed to limit computational processing. Motivated by the general lack of emphasis given to the understanding of audio rendering in virtual environments and games, we have started investigating multimodal (audiovisual) interactions within such virtual environments. Our work has shown that sound can directly affect visual fidelity perception and task performance within a virtual environment. These effects can be very individualized, whereby the influence of sound is dependent on various individual factors including musical listening preferences, suggesting the importance of individualizing the virtual environment to each user. This chapter begins with an overview of virtual environments and serious gaming's open problems, with an emphasis on fidelity, and multimodal interactions, and the implications that these may have on performance and computational requirements. A detailed summary of our own prior work will be provided along with insight and suggestions that may guide designers and developers of serious games and virtual

B. Kapralos (✉)
Faculty of Business and Information Technology,
University of Ontario Institute of Technology, Oshawa, ON, Canada
e-mail: bill.kapralos@uoit.ca

F. Moussa
Division of Cardiac and Vascular Surgery, Schulich Heart Centre, Sunnybrook Health Sciences Centre, Toronto, ON, Canada

K. Collins
The Games Institute, University of Waterloo, Waterloo, ON, Canada

A. Dubrowski
Divisions of Emergency Medicine and Pediatrics, Faculty of Medicine, Memorial University, St. John's, NL, Canada

P. Wouters, H. van Oostendorp (eds.), *Instructional Techniques to Facilitate Learning and Motivation of Serious Games*, Advances in Game-Based Learning, DOI 10.1007/978-3-319-39298-1_5

learning environments in general. Although the chapter is contextualized in the use of serious games in health professions education, the information provided is generalizable across a variety of domains.

Keywords Serious gaming • Virtual simulation • Fidelity • Realism • Audiovisual cue interaction, multimodal interaction

5.1 Introduction

One of the prevailing arguments for using physical simulations in health professions education is their ability to engage the trainee in the active accumulation of knowledge, skills, and attitudes by doing. However, physical simulation environments are costly to build and maintain (Cook et al., 2013; Isaranuwatchai, Brydges, Carnahan, Backstein, & Dubrowski, 2014). These costs include the purchase of equipment, maintenance, and costs related to human operation (teachers/educators, technologists, and other supportive staff). Logistical issues include ensuring trainees are released from their clinical duties to take part in training within the hours of operations of the training centers, ensuring adequate staffing and supervision.

The rising popularity of video games has seen a recent push toward the application of video game-based technologies to teaching and learning. Serious games, defined as games whose primary purpose is education and training as opposed to solely entertainment, take advantage of video games to motivate and engage players/learners for a specific purpose. With respect to students, this strong engagement has been associated with academic achievement (Shute, Ventura, Bauer, & Zapata-Rivera, 2009). In addition to promoting learning via interaction, serious games allow users to experience situations that are difficult (even impossible) to achieve in reality due to factors such as cost, time, and safety concerns (Squire & Jenkins, 2003). Recent hardware and computational advancements are providing developers the opportunity to develop applications that employ a high level of fidelity and novel interaction techniques using off-the-shelf consumer-level hardware and devices. Moreover, devices such as the Microsoft Kinect motion sensing video sensor allow users to interact with their application using a natural user interface that employs gestures thus removing the game controller and its inherent limitations.

With respect to health professions education, serious games provide medical trainees the opportunity to acquire both cognitive and technical skills outside of the medical environment in an engaging and cost-effective manner. They allow trainees the opportunity to train until they reach a specific competency level, thus better preparing them before exposure to live patients or before entering physical training centers which, as described earlier, are costly to maintain/operate and whose availability may be limited. Serious gaming can potentially bridge the simulation and gaming worlds by harnessing the educational value of technology-enhanced simulation to teach specific technical or cognitive skills to learners (not only the trainees but patients too), alongside the motivational, interactive, and engaging benefits inherent in games (de Ribaupierre et al., 2014).[1]

Despite the many benefits associated with serious games and their growing popularity, there are a number of issues related to both their development and integration within a curriculum that should be addressed before they become part of the mainstream teaching curriculum. As will be detailed later in this chapter, one of the issues pertains to fidelity; that is, how realistic the virtual environment that the serious game is centered on must be in order to ensure effective learning, while another issue pertains to multimodal interactions. The perception of fidelity is influenced by multimodal interactions, which has potentially significant implications for designers and developers of virtual environments, given that with our current technology, we cannot faithfully recreate a real-world scenario with equal consideration of all of the senses. Thus, we can introduce *perceptual-based rendering*, individualized to each user, to combat some of the need for higher fidelity.

We have recently formed an interdisciplinary team of researchers comprised of experts in computer science/game development, engineering, medicine/healthcare, and education to investigate such questions within a specific focus on health professions education. The overarching aim of this team is to develop a greater understanding of fidelity, multimodal interactions, perceptual-based rendering, user-specific factors, and their effect on learning with the ultimate goal of developing more effective serious games. Through a series of user studies, our work to date has methodically examined the direct effect of sound on the perception of visual fidelity (the degree to which visual features in the virtual environment conform to visual features in the real environment (Lee, Rincon, Meyer, Höllerer, & Bowman, 2013) and its relationship to task performance. Although this series of experiments has shown a strong influence of sound on visual fidelity perception and task performance, results have also shown strong individual effects, whereby the influence of sound is dependent on various personal, individual factors. This conflicting result has led to our current effort examining the customization of serious games to each user through a calibration process.

5.2 Prior Research on Fidelity

As described earlier, despite the benefits of serious games, there are open, fundamental issues related to fidelity, and multimodal interactions that require further investigation. Tashiro and Dunlap (2007) developed a typology of serious games for

[1] A complete discussion of the use of serious gaming and game-based technologies for health professions education is beyond the scope of this chapter. However, an overview of studies published focusing on serious games to teach some aspects of healthcare training is provided by de Ribaupierre et al. (2014) while an overview of serious gaming and virtual simulation and their application for surgical education and training is provided by Kapralos, Moussa, and Dubrowski (2014).

healthcare education and explored the strengths and limitations of serious games for improving clinical judgment. They identified seven areas that require research and improvements for the effective development of serious games one of which is the impact of fidelity on learning. In the context of serious games, fidelity denotes the extent to which the appearance and/or behavior of the simulation matches the appearance and behavior of the real system (Hays & Singer, 1989). Ker and Bradley (2010) divide fidelity into two components: (i) psychological fidelity, and (ii) physical fidelity. Psychological fidelity denotes the degree that the skills inherent in the real task being simulated are captured within the simulation (Ker & Bradley, 2010). This is related to cognitive fidelity, described with respect to placing users into realistic situations such that any decisions they make are based on what they would really know and expect (Chandler, Anthony, & Klinger, 2009). Cognitive fidelity has also been described as the extent to which the training captures the actual deep structure of the task (Veinott et al., 2014). Although cognitive fidelity is difficult to measure, within a high-fidelity virtual training environment, the cognitive demands of the training are similar to the cognitive demands of the real-world task being simulated (Gopher, 2006). Physical fidelity covers the degree of similarity between the training situation and the operational situation which is simulated (Hays & Singer, 1989; Ker & Bradley, 2010). Physical fidelity can be further divided into equipment fidelity that denotes the degree that the simulation replicates reality and environmental fidelity that denotes the degree that the simulation replicates the sensory cues (Ker & Bradley, 2010). Fidelity can also be described with respect to a level of realism (LoR) which Chalmers and Debattista (2009) define as the physical accuracy of a virtual environment required to achieve a one-to-one mapping of an experience in the virtual environment with the same experience in the real world. Without an appropriate LoR, users may adopt a different task strategy in the virtual world than in the real world (Chalmers & Debattista, 2009).

Further complicating our definitions of fidelity, McMahan (2011) classifies fidelity into three major components: (i) display, (ii) interaction, and (iii) simulation fidelity. Display fidelity is defined by how accurately the real-world environment is replicated, interaction fidelity is defined by how accurately the interactions performed in the real world are replicated, and simulation fidelity describes how faithfully the environment and objects seen in the real world are replicated (Lee et al., 2013). Display fidelity (ideally) encompasses all of the senses, although generally emphasis has been placed on recreating the visual scene, and, to a lesser degree, the acoustic scene. With the currently available technology, a simulated environment will not appear as realistic as the real world and therefore, such an environment will typically have a lower level of visual fidelity (Lee et al., 2013). Visual fidelity is difficult to describe precisely, and visual fidelity perception has been measured with respect to various rendering parameters including variations in shadow softness (sharpness of the shadow edges), smoothness of the surface, lighting, geometry, and texture quality. A study conducted by Rademacher, Lengyel, Cutrell, and Whitted (2001) demonstrated that both soft shadows and the type of surface affect the perception of visual fidelity. Slater, Khanna, Mortensen, and Yu (2009) examined the differences in visual fidelity of a scene rendered by real-time ray tracing (higher visual fidelity) vs. real-time ray casting (lower visual fidelity) and demonstrated that the real-time ray-tracing rendered scene led to higher levels of presence ((the psychological) sense of being in the virtual environment), indicating that higher visual fidelity does in fact affect presence (or what might also be called immersion). Ray tracing and ray casting are two methods that add realism to a rendered image by including variations in shade, color intensity, and shadows that would be produced by having one or more light sources in the scene.

As evidenced from the discussion earlier, it should be noted that fidelity in general is complex, encompassing various aspects of a simulation activity including the sensory modalities (visual, auditory, olfactory, and haptics), learning objectives, and task demands and therefore, the term can convey diverse meanings (Cook et al., 2013). For the remainder of this chapter, and as discussed relative to the series of experiments that we previously conducted, unless specified otherwise, fidelity will be used here to denote the level of realism primarily with respect to the visual scene.

Fidelity can impact learning through what is known as *transfer*. Transfer can be defined as the application of knowledge, skills, and attitudes acquired during training to the environment in which they are normally used (Muchinsky, 1989). Generally, it is assumed that the closer the context of learning to the context of practice, the better the learning and thus the greater the transfer (Godden & Baddeley, 1975). Designers and developers of immersive 3D virtual environments including serious games, typically aim to remain faithful in their representation of the real environment, striving for high fidelity. In the real world our senses are constantly exposed to stimuli from multiple sensory modalities (visual, auditory, vestibular, olfactory, and haptic) and although the process is not exactly understood, we are able to integrate/process this multisensory information and acquire knowledge of multisensory objects (Seitz, van Wassenhove, & Shams, 2007). However, despite the great computing hardware advances that we have recently experienced, particularly with respect to graphics rendering, real-time highly realistic rendering of complex environments across all modalities (i.e., a high degree of environmental fidelity) is still not feasible (Hulusic et al., 2012). In addition, there are other potential issues with striving for high-fidelity, multimodal virtual environments. More specifically, recent evidence suggests that high-fidelity simulation does not always lead to greater learning and thus it remains unclear if high fidelity is actually *needed* to maximize learning (Norman, Dore, & Grierson, 2012). Furthermore, striving for high-fidelity virtual environments can burden our computational resources (particularly with portable computing devices), increase the probability of lag and subsequent discomfort and simulator sickness (Blascovich & Bailenson, 2011), and lead to increased development costs.

In the real world, visuals are generally accompanied by auditory cues and the interaction between the two can affect our perception of fidelity of both the visual and auditory scene. *Perceptual-based rendering*—whereby the rendering parameters (typically of the visuals/graphics) are adjusted based on our perceptual system—is often employed to limit computational processing. One aspect of perceptual-based rendering is the use of selective rendering algorithms that allow known perceptual limitations to be translated into tolerances when rendering (Chalmers & Debattista, 2009).

Although prior work has demonstrated a strong influence between sound and the visual scene, the following question can still be posed: what, if any, consequences does perceptual-based rendering have within a serious gaming environment where the end result is the transfer of knowledge and/or skills development?

5.3 Related Work

A large number of studies have examined the role of fidelity and the interactions between the various parts of fidelity in both physical- and virtual-based simulation and serious games and its effect on transfer. For example, Chandler et al. (2009) described a case whereby soldiers who trained using a high physical fidelity simulation failed to function adequately during their deployment due to the deep structure differences between the training and reality. McMahan (2011) conducted a study that employed a six-sided Cave Automatic Virtual Environment (CAVE), visualization system that systematically evaluated the effects of very high and very low levels of display and interaction fidelity on the user experience within a first-person-shooter game. They found that display and interaction fidelity had significant positive effects on presence, engagement, and usability. They also suggest that designers and developers of serious games concerned with achieving high levels of presence, engagement, or usability should employ high levels of both display fidelity and interaction fidelity for the best results. Veinott et al. (2014) examined the interaction of cognitive and visual fidelity of a 3D serious game that was designed to train decision making. Participants played one of four different versions of the game (low cognitive and low visual fidelity, low cognitive and high visual fidelity, low visual and high cognitive fidelity, high visual and cognitive fidelity). Experimental results revealed an interaction between the two types of fidelity, but no main effects on learning. Learning was greatest when visual and cognitive fidelity matched and learning was least when visual fidelity was low, and cognitive fidelity was high. Studies have also shown that adding contextually relevant stress (e.g., increased psychological fidelity) to the simulation thus better mimicking the conditions to be encountered in the real-world actual setting can increase transfer (Driskell, Johnston, & Salas, 2001; Morris, Hancock, & Shirkey, 2004). For the remainder of this section, emphasis will be placed on multimodal interactions and in particular, on our own prior work that examined audiovisual interactions and more specifically, the effect of auditory cues on visual fidelity perception and task performance within a virtual environment.[2]

[2]Greater details regarding fidelity in general are available elsewhere. For example, Alexander, Brunyé, Sidman, and Weil (2005) provide a review of studies on fidelity, immersion, presence, and the resulting effects on transfer in simulations and games. Greater details regarding the influence of sound over visual rendering and task performance are provided by Hulusic et al. (2012) while an overview of multimodal influences on visual perception is provided by Shams and Kim (2010).

Various studies have examined the perceptual aspects of audiovisual cue interaction and have concluded that the perception of visual fidelity can affect the perception of sound fidelity (quality) and vice versa (Storms & Zyda, 2000). Numerous studies have demonstrated that multimodal effects can be considerable, to the extent that large amounts of detail of one sense may be ignored in the presence of other sensory inputs. For example, it has been shown that sound can potentially attract part of the user's attention away from the visual stimuli and lead to a reduced cognitive processing of the visual cues (Mastoropoulou, Debattista, Chalmers, & Troscianco, 2005). Hulusic, Debattista, Aggarwal, and Chalmers (2011) showed that sound effects allowed slow animations to be perceived as smoother than fast animations and that the addition of footstep sound effects to visual-based walking animations increased the animation smoothness perception, while Bonneel, Suied, Viaud-Delmon, and Drettakis (2010) observed that visual level of detail is perceived to be higher as the auditory level of detail is increased. This in turn may allow us to increase the fidelity of the audio channel if the enhancement of visuals within a virtual environment is economically or technically limited (Larsson, Västjäll, & Kleiner, 2003).

Our prior work (nine experiments) has so far examined visual fidelity perception in the presence of various auditory conditions. Participants in our nine experiments were recruited from several universities in Ontario, Canada and included both graduate and undergraduate students and researchers from various disciplines including Computer Science, Game Development, and Health Sciences (Nursing in particular). The majority of the participants were students whose age ranged from 19 to 24 and the average number of participants for each experiment was 14. Our initial studies included simple static environments comprised of a single 2D image of a surgeon's head (a rendered 3D model). In the first study (Experiment 1), visual fidelity was defined with respect to the 3D model's polygon count (Rojas et al., 2011), and in the second study (Experiment 2), polygon count was kept constant and visual fidelity was defined with respect to the 3D model's texture resolution (Rojas et al., 2012). In computer graphics, a 3D model begins with a wireframe that is constructed by connecting a number of polygons (typically triangles). Objects with a large polygon count generally lead to a more faithful (and realistic) representation (albeit, they do require greater computational power to render) while those with low polygon counts can appear blocky (but require less computational power). Once the wireframe has been generated, texture mapping, whereby a texture map (i.e., an image, color, or computer generated graphic), is added to the wireframe to provide detail and surface texture to the object. The texture resolution refers to the resolution of the image mapped to the wireframe; the greater the resolution, the greater the surface detail and generally the greater the computational requirements. The visual stimuli used in Experiment 1 are provided in Fig. 5.1, while the visual stimuli used in Experiment 2 are provided Fig. 5.2. In both studies, participants were presented with the static visual (the rendering of the surgeon's head). A total of six visuals were considered, each differing with respect to polygon count or texture resolution depending on the experiment, and each visual was presented along with one of four auditory conditions: (i) no sound at all (silence), (ii) white noise,

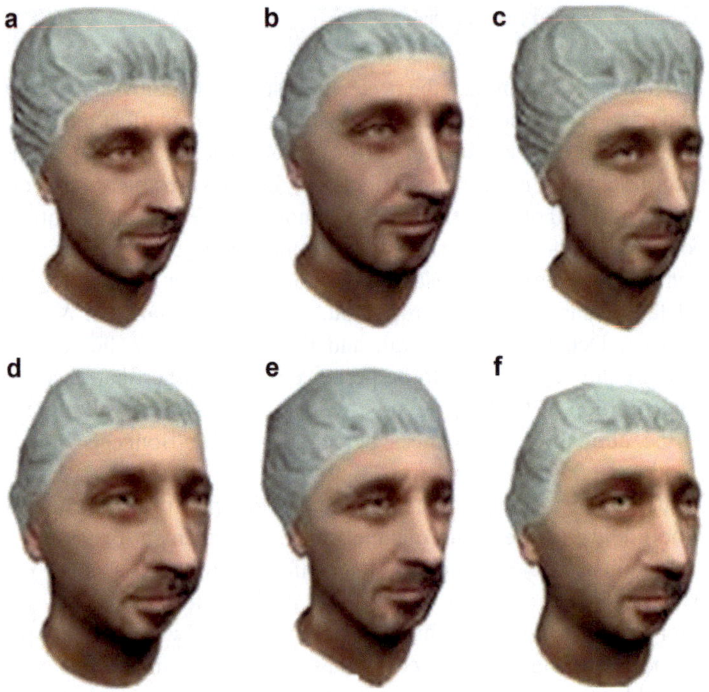

Fig. 5.1 The visual stimuli used in Experiments 1 and 3 consisting of a rendered 3D model of a surgeon's head, where each rendering varied with respect to polygon count defined as follows: (**a**) 17,440, (**b**) 13,440, (**c**) 1250, (**d**) 865, (**e**) 678, and (**f**) 548. Reprinted from Rojas et al. (2011)

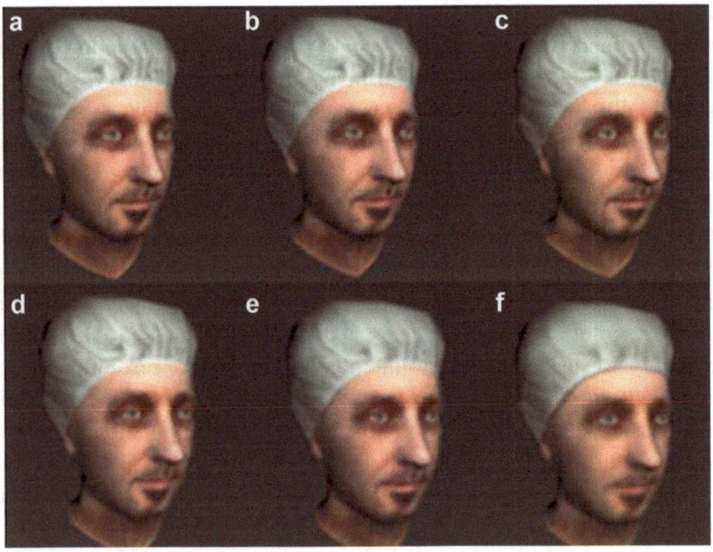

Fig. 5.2 The visual stimuli used in Experiments 2 and 4 consisting of a rendered 3D model of a surgeon's head, where each rendering varied with respect to texture resolution defined as follows: (**a**) 1024 × 1024, (**b**) 512 × 512, (**c**) 256 × 256, (**d**) 128 × 128, (**e**) 64 × 64, and (**f**) 32 × 32. Reprinted from Rojas et al. (2012)

(iii) classical music (Mozart), and (iv) heavy metal music (Megadeth). For each of the audiovisual presentations, the participants' task was to rate the fidelity of the visual on a scale from 1 to 7. With respect to Experiment 1, visual fidelity perception was, in general, greatest in the presence of classical music particularly when considering the renderings (visual stimuli) corresponding to higher polygon count. With respect to Experiment 2, sound consisting of white noise had very specific and detrimental effects on the perception of visual fidelity when considering high-fidelity renderings. In contrast to the study where visual fidelity was defined with respect to polygon count, the music-based auditory conditions (classical or heavy metal music) did not have any effect on the perception of visual fidelity when visual fidelity was defined with respect to texture resolution.

These two experiments were repeated with the identical visuals but now the visuals were presented in stereoscopic 3D, that is, with the illusion of depth created by delivering a separate view to the left and right eye in order to mimic human depth perception (Rojas et al., 2013). When visual fidelity was defined with respect to polygon count (Experiment 3), classical music led to an increase in visual fidelity perception while white noise had an attenuating effect on the perception of visual fidelity. However, both of these effects were evident for only the visual models whose polygon count was greater than 678 (i.e., sound did not have any effect on the two smallest polygon count models), indicating that there is a polygon count threshold after which the visual distinction is not great enough to be negatively influenced by sound. When visual fidelity was defined with respect to texture resolution (Experiment 4), both classical music and heavy metal music led to an increase in visual fidelity perception while white noise caused a decrease in visual fidelity perception.

The results of these initial four experiments have shown that sound can affect visual fidelity perception, and at times, the resulting effect can be substantial. However, the auditory conditions considered in these studies were noncontextual (no direct relationship) with respect to the visual scene. Two experiments were thus conducted to examine visual fidelity perception in the presence of contextual sounds, that is, sounds that had a causal relationship to the visual scene (Rojas et al., 2013; Rojas, Kapralos, Collins, & Dubrowski, 2014). As shown in Fig. 5.3, the visual stimuli consisted of six images of a surgeon holding a surgical drill, against a black background (similar to the visual stimuli employed in Experiments 2 and 4 as shown in Fig. 5.2 but with the addition of the surgeon's upper body and surgical drill). Visual fidelity was defined with respect to texture resolution; the polygon count of the 3D model was kept constant but the texture resolution of the surgeon and the drill was varied. The auditory conditions included the four noncontextual auditory conditions considered in the previous four experiments in addition to the following three contextual sounds: (i) operating room ambiance which included machines beeping, doctors and nurses talking; (ii) drill sound; and (iii) hospital operating room ambiance coupled (mixed) with the drill sound. The visuals remained static in both of the contextual auditory experiments but in the first of these two experiments (Experiment 5) the visuals were presented in nonstereoscopic 3D (i.e., the same visual was presented to the left and right eye and therefore, depth information arising from binocular cues was missing), while in the second of

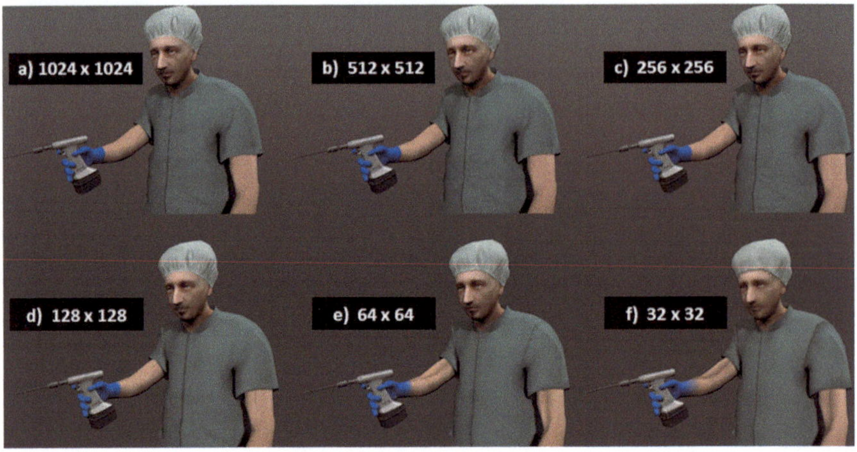

Fig. 5.3 The visual stimuli used in Experiments 5 and 6 consisting of a 3D model of a surgeon holding a drill. The six different texture resolutions ranged from 1024 × 1024 to 32 × 32 as shown in (**a**) to (**f**), respectively. Reprinted from Rojas et al. (2013)

these two experiments (Experiment 6), the visuals were presented in stereoscopic 3D. With nonstereoscopic 3D viewing, contextual auditory cues led to an increase in the perception of visual fidelity while noncontextual cues in the form of white noise led to a decrease in visual fidelity perception, particularly when considering the lower fidelity visuals (Rojas et al., 2014). However, the increase in visual fidelity perception was observed for only two of the three contextual auditory conditions and more specifically, for the operating room ambiance, and operating room ambiance + drill auditory conditions and not for the drill auditory condition despite the fact that the surgeon within the visual scene was holding a surgical drill. With respect to stereoscopic 3D viewing, white noise led to a decrease in visual fidelity perception across all of the visual conditions considered. However, none of the auditory conditions led to a statistically significant increase in visual fidelity perception (Rojas et al., 2013). That being said, none of the participants were surgeons or medical practitioners and may not have been familiar with an operating room and the sounds contained within an operating room. The notion of contextual auditory cues may be individualized and may depend on various individual factors including prior experience, expertise, and musical listening preferences.

The experiments described so far considered static visual environments where the visual scene (the 3D models presented to the participants) remained static. Three additional experiments were conducted to examine the effect of sound on visual fidelity perception and task performance within a dynamic virtual environment (a virtual surgical operating room from a serious game for total knee arthroplasty training (Cowan et al., 2010)), where the participants had to interact with the virtual operating room while completing a simple task. In each of the experiments, participants were required to complete a simple task within the virtual operating room. The task of the participants was to navigate through the virtual operating room from

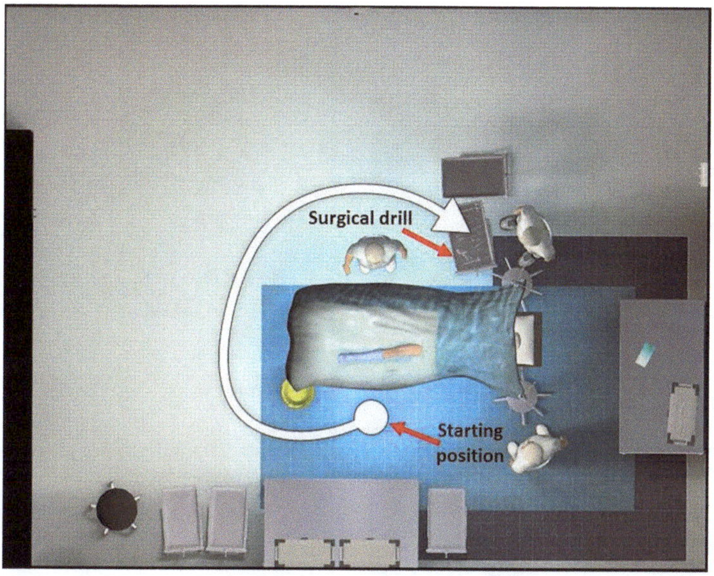

Fig. 5.4 Top-down view of the virtual operating room used in Experiments 7, 8, and 9. Reprinted from Cowan, Rojas, Kapralos, Moussa, and Dubrowski (2015)

their starting position to a point in the room which contained a tray with surgical instruments (see Fig. 5.4). Once they reached the tray, they were required to pick up a surgical drill (they had to navigate around a bed and a nonplayer character nurse to reach the tray that contained the surgical instruments). Upon picking up the surgical drill, participants were prompted to rank the visual scene with respect to their perceived visual fidelity.

The three experiments differed with respect to the audio and visual stimuli. In the first of these three experiments (Experiment 7), visual fidelity was defined with respect to the level of (consistent) blurring of the entire screen (level of blurring of the scene was used to approximate varying texture resolution), and the auditory cues consisted of the three contextual cues considered in the previous experiments (e.g., operating room ambiance and operating room ambiance coupled (mixed) with a surgical drill sound), in addition to white noise and no sound (Cowan et al., 2015). Sound (contextual and noncontextual) did not influence the perception of visual fidelity perception irrespective of the level of blurring. However, sound did impact performance (time to complete the required task). More specifically, white noise led to a large decrease in performance (increase in the amount of time required to complete the task) while the two contextual sounds considered improved performance (decrease in the amount of time required to complete task), across all levels of visual fidelity considered. In the second of these three experiments (Experiment 8) (Rojas et al., 2015), visual fidelity was defined with respect to several visual filtering effects including levels of cel shading. Cel shading is also known as toon shading (or cartoon shading) and is a popular 3D rendering technique used in games that

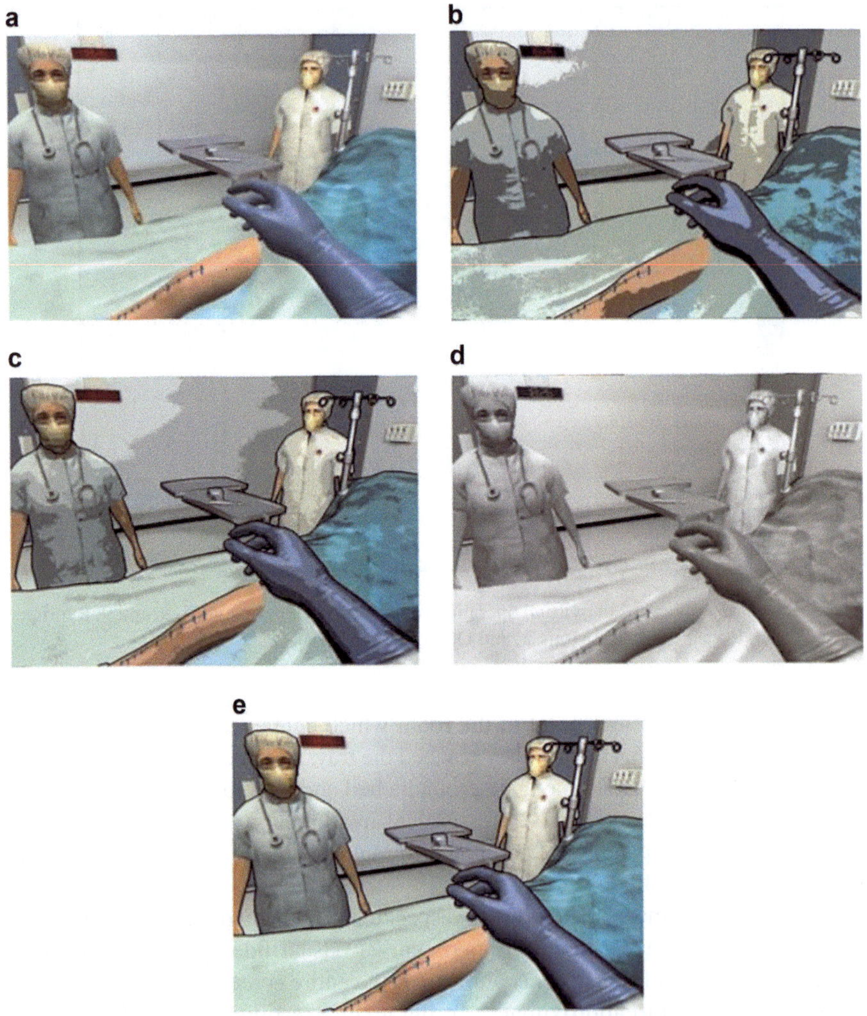

Fig. 5.5 The visual stimuli used in Experiment 8. (**a**) Original (unmodified), (**b**) 3-level cel shading, (**c**) 6-level cel shading, (**d**) outline, and (**e**) grayscale. Taken from Cowan et al. (2015)

attempts to recreate the look of traditional 2D animation with the use of a set of flat colors ("levels") that leads to a "cartoon look." The visual stimuli used in Experiment 8 are shown in Fig. 5.5. The original (unmodified) version is shown in Fig 5.5a. Two levels of cel shading were employed: (i) 3-level cel shading (i.e., color was divided into three discrete levels; see Fig. 5.5b), and (ii) 6-level cel shading (i.e., color was divided into six discrete levels; see Fig. 5.5c). An additional two visual effects (conditions) was also considered. More specifically, the grayscale effect removed all color from the visual scene (see Fig. 5.5d), and the outline effect highlighted the edges (outlines) of objects in the scene (see Fig. 5.5e).

The auditory conditions in this second experiment consisted of no sound, two contextual sounds (operating room ambiance mixed with a surgical drill sound, and operating room ambiance without the drill sound), and two noncontextual sounds (white noise and classical music (Mozart)). Here, white noise led to a decrease in visual fidelity perception but no other effects of sound on visual fidelity were observed. That being said, although not statistically significant, the auditory conditions that led to higher ratings of visual fidelity perception were those where the content (e.g., operating room ambiance and the operating room ambiance with surgical drill sound) was related to the context of the visual scenario that they were being presented with (e.g., a graphically rendered operating room). With respect to performance, the grayscale visual condition led to the greatest performance. Participants performed the task faster under the grayscale visual condition when compared to the other four visual conditions. Sound did not have any effect on task completion time for the other visual conditions. Finally, the last of these three experiments (Experiment 9) was identical to Experiment 7 but the sounds were spatialized by adding reverberation (i.e., echoes) to allow the user to perceive the sound as emanating from a particular position in three-dimensional space (Cowan, Rojas, Kapralos, Collins, & Dubrowski, 2013). For example, the surgical drill sound was spatialized such that it would be (ideally) perceived to be emanating from the position of the surgical drill. Contrary to all of the prior experiments, the presence of sound (spatial and nonspatial) did not have any effect on either visual fidelity perception or task completion time. That being said, only six participants took part in the experiment (in contrast to 12–18 for each of our previous experiments), thus the results require further investigation and is the focus of our future work.

5.3.1 Summary and Discussion of Our Experimental Results

As summarized in Table 5.1, the first six experiments considered static visuals that were presented in either stereoscopic 3D or nonstereoscopic 3D while, as summarized in Table 5.2, the last three experiments considered a dynamic virtual environment that required the user to interact with. Aside from the fact that white noise generally led to a decrease in the perception of visual fidelity and task performance, and classical music led to an increase in visual fidelity perception (the majority of the time) across all of the experiments considered, the results varied significantly across each of the experiments making it difficult to draw any firm conclusions or to provide any design guidelines and/or best practices. The visuals and many of the sounds considered in these experiments were medical in nature, yet many of the participants were students and although some were enrolled in Health Sciences-related programs, they had limited (if any), medical-based training and operating room exposure. We hypothesize that the variation seen across the results of these experiments is due to individual differences. In other words, the influence of sound on visual fidelity perception and task performance can be complex, individualized, and may be highly influenced by past experiences, and musical preferences, among other factors

Table 5.1 Summary of results for Experiments 1–6

Exp.	Visual stimuli	Auditory stimuli	Results (visual fidelity perception)
1	**Model**: Rendering of surgeon's head **Fidelity**: Polygon count **Stereoscopic 3D**: No	**Noncontextual**: (i) no sound, (ii) white noise, (iii) classical music, and (iv) heavy metal music	**Increase**: (i) Classical music—more pronounced with higher polygon count, and (ii) heavy metal music **Decrease**: White noise
2	**Model**: Rendering of surgeon's head **Fidelity**: Texture resolution **Stereoscopic 3D**: No		**Increase**: None **Decrease**: White noise
3	**Model**: Rendering of surgeon's head **Fidelity**: Polygon count **Stereoscopic 3D**: Yes		**Increase**: Classical music—evident only when polygon count was greater than 678 **Decrease**: White noise—evident only when polygon count was greater than 678
4	**Model**: Rendering of surgeon's head **Fidelity**: Texture resolution **Stereoscopic 3D**: Yes		**Increase**: (i) Classical music, and (ii) heavy metal music **Decrease**: (i) White noise
5	**Model**: Rendering of surgeon holding drill **Fidelity**: Texture resolution **Stereoscopic 3D**: No	**Noncontextual**: (i) no sound, (ii) white noise, (iii) classical music, and (iv) heavy metal music **Contextual**: (i) operating room ambiance, (ii) surgical drill, and (iii) operating room ambiance mixed with surgical drill	**Increase**: (i) operating room ambiance, and (ii) operating room ambiance + drill (both were contextual sounds) **Decrease**: White noise—more pronounced for lower realism visuals
6	**Model**: Rendering of surgeon holding drill **Fidelity**: Texture resolution **Stereoscopic 3D**: Yes		**Increase**: None **Decrease**: (i) White noise

(e.g., see Anyanwu, 2015; Avila, Furnham, & McClelland, 2012). As a result, we believe that a one-size-fits-all approach to perceptual-based rendering will not necessarily account for these individual differences and thus a more individualized approach is needed. Our future work includes repeating these dynamic task-based experiments (i.e., Experiments 7, 8, and 9) with a larger number of participants and with participants that have prior operating room exposure and experience (e.g., surgical residents, operating room nurses, surgeons, and anesthetists). Aside from allowing us to confirm or refute the preliminary results obtained in Experiment 9 (that considered only six participants), it will allow us to examine the influence that prior experience/exposure has with respect to contextual sounds and their influence on the visual scene.

Table 5.2 Summary results of Experiments 7–9

Exp.	Visual stimuli (fidelity)	Auditory stimuli	Results
7	**Model**: Rendering of an operating room **Fidelity**: Levels of blurring **Stereoscopic 3D**: No	**Noncontextual**: (i) no sound, and (ii) white noise **Contextual**: (i) operating room ambiance, (ii) surgical drill, and (iii) operating room ambiance mixed with surgical drill **Spatial sound**: No	**Visual fidelity perception**: No influence of sound (contextual and noncontextual) on the perception of visual fidelity perception irrespective of the level of blurring **Task performance**: (i) white noise led to a large decrease in performance across visual conditions, and (ii) both contextual sounds increased performance across all visual conditions
8	**Model**: Rendering of an operating room **Fidelity**: 3-level cel shading, 6-level cel shading (i.e., color was divided, outline, and grayscale **Stereoscopic 3D**: No	**Noncontextual**: (i) no sound, (ii) white noise, and (iii) classical music (Mozart) **Contextual**: (i) operating room ambiance, and ii) operating room ambiance mixed with surgical drill **Spatial sound**: No	**Visual fidelity perception**: (i) White noise led to a decrease in visual fidelity perception across all visual conditions, and (ii) although not significant, contextual auditory conditions led to higher ratings of visual fidelity perception **Task performance**: (i) Grayscale visual condition led to increased performance, (ii) white noise led to decreased performance across all visual conditions
9	**Model**: Rendering of an operating room **Fidelity**: Levels of blurring **Stereoscopic 3D**: No	**Noncontextual**: (i) no sound, and (ii) white noise **Contextual**: (i) operating room ambiance, (ii) surgical drill, and (iii) operating room ambiance mixed with surgical drill **Spatial sound**: Yes	**Visual fidelity perception**: No effects **Task performance**: No effects

The variation observed across both the experimental results and among participants within the individual experiments, and the potential impact this may have on perceptual-based rendering and learning has motivated our most recent work examining the customization of the serious gaming interface, and the fidelity of both the sound and graphics (and any perceptual-based rendering) presented to the user of a serious game through a calibration process that they complete before beginning the game or the simulation (Kapralos, Shewaga, & Ng, 2014). Although the work is ongoing and preliminary, the calibration method includes the use of a brief questionnaire that users complete prior to beginning the serious game/virtual simulation followed by an interactive calibration game that enables the user to choose the optimal audiovisual fidelity settings dynamically by playing a game. In this calibration game, the user is presented with a split screen with the same game running in each window but under different fidelity settings and a single background sound. In a manner similar to standard testing methodologies employed by optometrists to determine the optimum properties of corrective lenses in order to overcome a variety of visual deficiencies, the user chooses the window they prefer and the audiovisual fidelity of the game running in the other window will change. This process repeats over a number of cycles until the optimal fidelity level is reached. Although greater work remains, preliminary results indicate that the calibration game is both engaging, enjoyable, and fun although it still remains to be determined whether having the user customize the fidelity of the visual scene that they will be presented with will lead to more effective serious games and more specifically, to greater transfer.

5.4 Conclusions, General Implications, and Guidelines

Recent hardware and computational advancements are providing designers and developers of serious games the opportunity to develop applications that employ a high level of fidelity and novel interaction techniques using off-the-shelf consumer-level hardware and devices. Although designers and developers of serious games (and virtual simulations in general) typically take advantage of the computational advancements and strive for high visual fidelity environments, past research has been inconclusive with respect to the effect of visual fidelity on learning. More specifically, some studies have shown that greater visual fidelity is better (Cooke & Shope, 2004), while on the contrary, other studies suggest that improved learning outcomes result with lower visual fidelity environments that maintain the deep cognitive structure of the task in question (Veinott et al., 2014). However, as shown, fidelity in general is complex; it can be decomposed into a large number of components including physical fidelity, psychological fidelity, environmental fidelity, cognitive fidelity and it can be affected by a large number of factors including a learner's prior experience and expertise. With respect to serious games and virtual simulations, the term fidelity has generally been implied to graphical rendering given the emphasis placed on the visual scene. Yet, there is far more to a virtual learning environment than the graphical scene that can both positively and negatively affect a user's perception of the visual scene.

There are many factors besides fidelity that influence the issues of cognitive load and learning that make teasing out the important parameters of audiovisual fidelity that impact learning difficult. Fidelity may be more or less important to specific types of learning or specific types of activities. For instance, the depth of cognitive engagement required to complete a task may be influenced by fidelity more or less than a more physically demanding task. Moreover, as described, these issues may be individualistically determined, and thus experiments need to consider the individual preferences for learning. For example, some people may prefer background music to study, while others may prefer silence (e.g., see Furnham & Strbac, 2002). Another important consideration is the learner's level of expertise which will directly influence the cognitive load that can be placed on them (Chandler et al., 2009). More specifically, virtual environments with a high level of visual fidelity typically result in increased cognitive load and since a more expert learner can typically accommodate a higher cognitive load, higher visual fidelity virtual environments may be better suited to more expert learners (Chandler et al., 2009). This can be explained by Treisman and Riley's (1969) sensory integration model and cognitive load theory. The sensory integration model states that the central nervous system integrates the various sensory inputs that it is constantly receiving to arrive at a coherent representation of the external world. The model hypothesizes that information from all of the stimuli presented to the senses at any given time enters a sensory buffer and one of the inputs is then selected on the basis of its physical characteristics for further processing by being allowed to pass through a filter. In other words, given our limited processing capacity, we cannot attend to all of the information available to us at any one point in time and, therefore, the information presented to us is passed through a filter to prevent the system from overloading. This can further explained by the cognitive load theory (Paas, Renkl, & Sweller, 2003). Cognitive load theory contends that during complex learning/training activities, the amount of information and interactions that must be processed simultaneously can either underload or overload the finite amount of working memory (Paas et al., 2003). Here, the learner's expertise should not be ignored. More specifically, according to Sweller, Ayres, Kalyuga, and Chandler (2003), many of the cognitive load theory (CLT) effects that can be used to recommend instructional designs are only applicable to novice learners and as expertise increases, these effects begin to disappear, ultimately reversing. This is known as the expertise reversal effect and as a result, instructional techniques that are effective with novice learners can lose their effectiveness and even have negative consequences when applied to more experienced learners (Sweller et al., 2003). Yet another explanation is provided by Mayer and Moreno (2003) with respect to multimedia learning, that is, learning from "words and pictures." In such a situation, the processing demands evoked by the learning task may exceed the processing capacity of the cognitive system—a situation they refer to as cognitive overload.

Whether or not a secondary modality plays an important role in learning could also influence the impact of higher fidelity in one or more modes. In our own series of experiments that involved a dynamic virtual environment where participants were required to complete a task (Experiments 7, 8, and 9), they focused primarily in completing the task (navigating the virtual environment to select a surgical drill) and therefore, the visual input was the most relevant cue for them to complete the task. However, they were also presented with auditory input. Given that the participants were not familiar with the sounds and the objects of an operating room, according to cognitive load theory, their cognitive load may have been overloaded and therefore the audio cues were ignored given that they (the auditory cues) were not necessarily required to navigate the virtual operating room to pick up the surgical drill. However, the sound of the drill, for instance on the skull, is an important sound to a surgeon who, through the changes in sound during drilling, knows when the drill has broken through the bone/skull into the blood-brain barrier. The question then becomes how much visual fidelity is needed to conjure the required cognitive performance? (Chandler et al., 2009). This question has important implications aside from learning. More specifically, lower visual fidelity environments are typically cheaper and quicker to produce with respect to cognitive performance, may be just as effective as a higher fidelity version (Chandler et al., 2009).

Although so far fidelity has been considered independently or with respect to multimodal interactions, it can also influence various other important parameters. According to Slater and Wilbur (1997), immersion is a description of a technology and describes the extent to which the computer displays are capable of delivering an inclusive, extensive, surrounding, and vivid illusion of reality to the senses of a human participant. The influence of visual fidelity (realism) on immersion has been studied extensively in an attempt to establish whether a greater level of fidelity does indeed complement (increase) immersion (Netepezuk, 2013). Results have so far been mixed with several studies finding a definite effect on immersion and engrossment of the player (Bracken & Skalski, 2006) while others have found a lack of effects (Cheng & Cairns, 2005). This is an area that requires further investigation.

Despite the growing popularity of serious games in medical education and wide variety of other areas, there are many examples of ineffective serious games, that is, serious games that provide little, if any, educational value. This has been highlighted in a series of studies that have examined the effectiveness of virtual simulations and game-based learning, prompting some to challenge the usefulness of game-based learning in general (Cannon-Bowers, 2006; Gosen & Washbush, 2004; Hays, 2005). Hays (2005) has attributed this to the lack of proper instructional design and suggests the need to embark on further studies to evaluate the value of instructional design in the game development process (Hays, 2005). To develop effective serious games, care must be taken to ensure that they are properly designed to meet their intended goals (Becker & Parker, 2012). Part of this includes conducting a proper and thorough needs and task analysis; we believe that during this stage, the question of fidelity should be addressed by carefully considering the outcomes/objectives, target audience, and budget, among others factors. Careful consideration should also be given to cognitive load and more specifically, multimedia-based instruction in general should be designed such that any unnecessary cognitive load is minimized so that the occurrence of cognitive overload scenario is therefore minimized (Mayer & Moreno, 2003). However, minimizing a cognitive overload scenario is not a trivial task for instructional designers given that meaningful learning (i.e., a deep understanding of the material, and the ability to apply what was taught to new situations) may require a large amount of cognitive resources that often exceeds the learner's cognitive capacity (Mayer & Moreno, 2003).

With respect to simulation-based medical education, after a critical evaluation of historical and contemporary research on simulation-based medical education over a 40-year span, McGaghie, Issenberg, Petrusa, and Scalese (2010) proposed 12 features and best practices of simulation-based education (along with current gaps in understanding for each feature), that simulation developers and educators should have knowledge to ensure maximum educational benefit when employing medical-based simulation technology. The 12 features and best practices are as follows: (i) feedback, (ii) deliberate practice, (iii) curriculum integration, (iv) outcome measurement, (v) simulation fidelity, (vi) skill acquisition and maintenance, (vii) mastery learning, (viii) transfer to practice, (ix) team training, (x) high-stakes testing, (xi) instructor training, and (xii) educational and professional context. A complete discussion regarding these 12 features is beyond the scope of this work, with respect to fidelity, they suggest that a close match of education goals with simulation tools is required. For example, when considering basic procedural skills such as suturing, low-fidelity tools can be used (e.g., task trainers that mimic body) whereas more complex clinical events such as team responses to simulated hospital "codes" require training more sophisticated training tools (e.g., lifelike full-body manikins with computer-driven physiological features (e.g. heart rate, blood pressure). That being said, McGaghie et al. (2010), also acknowledge the knowledge gap regarding just how much fidelity is too much or enough with little prior published work regarding the effectiveness of high-fidelity simulations in medical education. Despite the difficulties associated with designing and developing serious games, there is a lack of any widely accepted best practices available to designers and developers of serious games to assist in the process. Although developing a list of best practices for serious game development is not a trivial task and will require great effort and further studies, a good starting point is a careful consideration of how the list of 12 features and best practices proposed by McGaghie et al. (2010) can be applied (and perhaps mapped) to serious games. We believe that a widely accepted list of best practices that guides designers and developers throughout the development process and addresses the issue of fidelity will be of great value and assistance to designers and developers of serious games.

As described, we also believe that a one-size-fits-all approach is not ideal. Rather, it is important to allow for customization of the user interface, perceptual-based rendering, and the various simulation parameters that relate to fidelity. This area requires greater investigation but current work (including our own that is examining the use of a game-based calibration method) in this domain is promising. Finally, given the individual differences regarding the effect of sound on visual fidelity perception and task performance, serious games and virtual simulations can be used to help trainees learn how to perform under the presence of potentially distracting sounds which, in many situations, characterize the real-world environment and cannot be eliminated. Virtual simulations and serious games can be used to explicitly acquaint trainees with distracting sounds common in a real-world scenario being trained for and thus minimize any negative effects when distracting sounds are encountered in the real world.

Ultimately, here we have shown that high fidelity in one modality does not necessarily impact learning. Serious game designers may be able to achieve design goals with lower fidelity (and lower cost). However, as shown, the degree that the lower fidelity may impact learning is yet to be determined in the context of multimodal interactions and individual preferences.

Acknowledgments This work was supported in part by the *Natural Sciences and Engineering Research Council of Canada* (NSERC), the *Social Sciences and Humanities Research Council of Canada* (SSHRC), *Interactive & Multi-Modal Experience Research Syndicate* (IMMERSe) initiative, and the Canadian *Network of Centres of Excellence* (NCE), *Graphics, Animation, and New Media* (GRAND) initiative.

References

Alexander, A. L., Brunyé, T., Sidman, J., & Weil, S. A. (2005). *From gaming to training: A review of studies on fidelity, immersion, presence, and buy-in and their effects on transfer in PC-based simulations and games*. DARWARS Training Impact Group, November. Retrieved from CiteSeer.

Anyanwu, E. G. (2015). Background music in the dissection laboratory. Impact on stress associated with the dissection experience. *Advances in Physiology Education, 39*(2), 96–10.

Avila, C., Furnham, A., & McClelland, A. (2012). The influence of distracting familiar vocal music on cognitive performance of introverts and extraverts. *Psychology of Music, 40*(1), 84–93.

Becker, K., & Parker, J. (2012). *The guide to computer simulations and games*. Indianapolis, IN: Wiley.

Blascovich, J., & Bailenson, J. (2011). *Infinite reality*. New York, NY: Harper Collins.

Bonneel, N., Suied, C., Viaud-Delmon, I., & Drettakis, G. (2010). Bimodal perception of audio-visual material properties for virtual environments. *ACM Transactions on Applied Perception, 7*(1), 1–16.

Bracken, C., & Skalski, P. (2006, August). *Presence and video games: The impact of image quality and skill level*. Paper presented at the 9th Annual International Workshop on Presence. Retrieved from http://ispr.info/presence-conferences/previous-conferences/presence-2006/

Cannon-Bowers, J. (2006, March). The state of gaming and *simulation*. Paper presented at the Training 2006 Conference and Expo.

Chalmers, A., & Debattista, K. (2009, March). *Levels of realism for serious games*. Paper presented at the 2009 Conference in Games and Virtual Worlds for Serious Applications. doi:10.1109/VS-GAMES.2009.43.

Chandler, T., Anthony, M., & Klinger, D. (2009, June). *Applying high cognitive vs, high physical fidelity within serious games*. Paper presented at the Interservice/Industry Training, Simulation, and Education Conference.

Cheng, K., & Cairns, P. A. (2005, April). *Behaviour, realism and immersion in games*. Paper presented at the CHI '05 Extended Abstracts on Human Factors in Computing Systems. doi:10.1145/1056808.1056894.

Cook, D. A., Hamstra, S. J., Brydges, R., Zendejas, B., Szostek, J. H., Wang, A. T., et al. (2013). Comparative effectiveness of instructional design features in simulation-based education: Systematic review and meta-analysis. *Medical Teacher, 35*(1), e867–e898.

Cooke, N. J., & Shope, S. M. (2004). Designing a synthetic task environment. In S. G. Schiflett, L. R. Elliott, E. Salas, & M. D. Coovert (Eds.), *Scaled worlds: Development, validation, and application* (pp. 263–278). Surry, England: Ashgate.

Cowan, B., Rojas, D., Kapralos, B., Collins, K., & Dubrowski, A. (2013, June). *Spatial sound and its effect on visual quality perception and task performance within a virtual environment*. Paper presented at the 21st International Congress on Acoustics. doi:http://dx.doi.org/10.1121/1.4798377.

Cowan, B., Rojas, D., Kapralos, B., Moussa, F., & Dubrowski, A. (2015). Effects of sound on visual realism perception and task performance. *Visual Computer, 31*(9), 1207–1216.

Cowan, B., Sabri, H., Kapralos, B., Porte, M., Backstein, D., Cristancho, S., et al. (2010). A serious game for total knee arthroplasty procedure education and training. *Journal of Cybertherapy and Rehabilitation, 3*(3), 285–298.

de Ribaupierre, S., Kapralos, B., Haji, F., Stroulia, E., Dubrowski, A., & Eagleson, R. (2014). Healthcare training enhancement through virtual reality and serious games. In M. Ma, C. Lakhmi, L. Jain, & P. Anderson (Eds.), *Virtual, augmented reality and serious games for healthcare* (pp. 9–27). Berlin, Germany: Springer.

Driskell, J. E., Johnston, J. H., & Salas, E. (2001). Does stress training generalize to novel settings? *Human Factors, 43*(1), 99–110.

Furnham, A., & Strbac, L. (2002). Music is as distracting as noise: The differential distraction of background music and noise on the cognitive test performance of introverts and extraverts. *Ergonomics, 45*(3), 203–217.

Godden, D. R., & Baddeley, A. D. (1975). Context dependent memory in two natural environments: On land and underwater. *British Journal of Psychology, 66*(3), 325–331.

Gopher, D. (2006). Emphasis change as a training protocol for high demand tasks. In A. Kramer, D. Wiegman, & A. Kirlik (Eds.), *Attention: From theory to practice* (pp. 209–224). New York, NY: Oxford Psychology Press.

Gosen, J., & Washbush, J. (2004). A review of scholarship on assessing experiential learning effectiveness. *Simulation & Gaming, 35*(2), 270–293.

Hays, R. T. (2005). *The effectiveness of instructional games: A literature review and discussion* (Technical Report 2005-004). Orlando, FL: Naval Air Warfare Center, Training Systems Division.

Hays, R. T., & Singer, M. (1989). *Simulation fidelity in training system design*. New York, NY: Springer.

Hulusic, V., Debattista, K., Aggarwal, V., & Chalmers, A. (2011). Maintaining frame rate perception in interactive environments by exploiting audio-visual cross-modal interaction. *Visual Computer, 27*(1), 57–66.

Hulusic, V., Harvey, C., Debattista, K., Tsingos, N., Walker, S., Howard, D., et al. (2012). Acoustic rendering and auditory-visual cross-modal perception and interaction. *Computer Graphics Forum, 31*(1), 102–131.

Isaranuwatchai, W., Brydges, R., Carnahan, H., Backstein, D., & Dubrowski, A. (2014). Comparing the cost-effectiveness of simulation modalities: A case study of peripheral intravenous catheterization training. *Advances in Health Sciences Education, 19*(2), 219–232.

Kapralos, B., Moussa, F., & Dubrowski, A. (2014). An overview of virtual simulations and serious games for surgical education and training. In M. Ma, C. Lakhmi, L. Jain, & P. Anderson (Eds.), *Virtual, augmented reality and serious games for healthcare* (pp. 289–306). Berlin, Germany: Springer.

Kapralos, B., Shewaga, R., & Ng, G. (2014). Serious games: Customizing the audio-visual interface. In R. Shumaker, & S. Lackey (Eds.), *Virtual, augmented and mixed reality. Virtual, augmented and mixed reality. Applications of virtual and augmented reality. Lecture notes in computer science* (Vol. 8526, pp. 190–199).

Ker, J., & Bradley, P. (2010). Simulation in medical education. In T. Swanwick (Ed.), *Understanding medical education: Evidence, theory and practice* (pp. 164–190). West Sussex, UK: Wiley-Blackwell.

Larsson, P., Västjäll, D., & Kleiner, M. (2003, April). *On the quality of experience: A multi-modal approach to perceptual ego-motion and sensed presence in virtual environments*. Paper presented at the First International Speech Communications Association Tutorial and Research Workshop on Auditory Quality of Systems. Retrieved from http://www.isca_speech.org/archive_open/aqs2003

Lee, C., Rincon, G. A., Meyer, G., Höllerer, T., & Bowman, D. A. (2013). The effects of visual realism on search tasks in mixed reality simulation. *IEEE Transactions on Visual Computing and Graphics, 19*(4), 547–556.

Mastoropoulou, G., Debattista, K., Chalmers, A., & Troscianco, T. (2005, August). *The influence of sound effects on the perceived smoothness of rendered animations*. Paper presented at the 2nd Symposium on Applied Perception in Graphics and Visualization. doi:10.1145/1080402.1080404.

Mayer, R. E., & Moreno, R. (2003). Nine ways to reduce cognitive load in multimedia learning. *Educational Psychology, 38*(1), 43–52.

McGaghie, W. C., Issenberg, S. B., Petrusa, E. R., & Scalese, R. J. (2010). A critical review of simulation-based medical education research: 2003–2009. *Medical Educator, 44*(1), 50–63.

McMahan, R. P. (2011). *Exploring the effects of higher-fidelity display and interaction for virtual reality games*. Doctoral dissertation. Retrieved from Digital Libraries and Archives (etd-12162011-140224).

Morris, C. S., Hancock, P. A., & Shirkey, E. C. (2004). Motivational effects of adding context relevant stress in PC-based game training. *Military Psychology, 16*(1), 135–147.

Muchinsky, P. M. (1989). *Psychology applied to work*. Summerfield, NC: Hypergraphic Press.

Netepezuk, D. W. (2013, July). *Immersion and realism in video games—The confused moniker of video game engrossment*. Paper presented at the 18th International Conference on Computer Games. doi:10.1109/CGames.2013.6632613.

Norman, G., Dore, K., & Grierson, L. (2012). The minimal relationship between simulation fidelity and transfer of learning. *Medical Educator, 46*(7), 636–647.

Paas, F., Renkl, A., & Sweller, J. (2003). Cognitive load theory and instructional design: Recent developments. *Educational Psychologist, 38*(1), 1–4.

Rademacher, P., Lengyel, J., Cutrell, E., & Whitted, T. (2001). Measuring the perception of visual realism in images. In S. J. Gortler, & K. Myszkowski (Eds.), *Proceedings of the 12th Eurographics Workshop on Rendering Techniques* (pp. 235–248): London, UK: Springer.

Rojas, D., Cowan, B., Kapralos, B., Collins, K., & Dubrowski, A. (2015, June). *The effect of sound on visual quality perception and task completion time in a cel-shaded serious gaming virtual environment*. Paper presented at the 7th IEEE International Workshop on Quality of Multimedia Experience. doi:10.1109/QoMEX.2015.7148136.

Rojas, D., Kapralos, B., Collins, K., & Dubrowski, A. (2014). The effect of contextual sound cues on visual fidelity perception. *Studies in Health Technology and Informatics, 196*, 346–352.

Rojas, D., Kapralos, B., Cristancho, S., Collins, K., Hogue, A., Conati, C., et al. (2012). Developing effective serious games: The effect of background sound on visual fidelity perception with varying texture resolution. *Studies in Health Technology and Informatics, 173*, 386–392.

Rojas, D., Kapralos, B., Crsitancho, S., Collins, K., Conati, C., & Dubrowski, A. (2011, September). *The effect of background sound on visual fidelity perception*. Paper presented at the ACM Audio Mostly 2011 Conference—6th Conference on Interaction with Sound. doi:10.1145/2095667.2095675.

Rojas, D., Kapralos, B., Hogue, A., Collins, K., Nacke, L., Crsitancho, C., et al. (2013). The effect of ambient auditory conditions on visual fidelity perception in stereoscopic 3D. *IEEE Transactions on Systems, Man, and Cybernetics Part B: Cybernetics, 43*(6), 1572–1583.

Seitz, A. R., van Wassenhove, V., & Shams, L. (2007). Simultaneous and independent acquisition of multisensory and unisensory associations. *Perception, 36*(10), 1445–1453.

Shams, L., & Kim, R. (2010). Crossmodal influences on visual perception. *Physics Life Reviews, 7*(3), 295–298.

Shute, V. J., Ventura, M., Bauer, M., & Zapata-Rivera, D. (2009). Melding the power of serious games and embedded assessment to monitor and foster learning. In U. Ritterfeld, M. Cody, & P. Vorderer (Eds.), *Serious games: Mechanisms and effects* (pp. 295–321). New York, NY: Routedle.

Slater, M., Khanna, P., Mortensen, J., & Yu, I. (2009). Visual realism enhances realistic response in an immersive virtual environment. *IEEE Computer Graphics and Applications, 29*(3), 76–84.

Slater, M., & Wilbur, M. (1997). A framework for immersive virtual environments (FIVE): Speculations on the role of presence in virtual environments. *Presence Teleoperators and Virtual Environments, 6*(6), 603–616.

Squire, K., & Jenkins, H. (2003). Harnessing the power of games in education. *Insight, 3*, 5–33.

Storms, S. L., & Zyda, M. J. (2000). Interactions in perceived quality of auditory-visual displays. *Presence Teleoperators and Virtual Environments, 9*(6), 557–580.

Sweller, J., Ayres, P. L., Kalyuga, S., & Chandler, P. A. (2003). The expertise reversal effect. *Educational Psychologist, 38*(1), 23–31.

Tashiro, J., & Dunlap, D. (2007, November). *The impact of realism on learning engagement in educational games*. Paper presented at the ACM 2007 Conference on Future Play. doi:10.1145/1328202.1328223.

Treisman, A. M., & Riley, J. G. (1969). Is selective attention selective perception or selective response? A further test. *Journal of Experimental Psychology, 79*(1), 27–34.

Veinott, E. S., Perleman, B., Polander, E., Leonard, J., Berry, G., Catrambone, R., et al. (2014, October). *Is more information better? Examining the effects of visual and cognitive fidelity on learning in a serious video game*. Paper presented at the 2014 IEEE Games Entertainment and Media Conference. doi:10.1109/GEM.2014.7048105.

Chapter 6
Narration-Based Techniques to Facilitate Game-Based Learning

Herre van Oostendorp and Pieter Wouters

Abstract In this chapter, we discuss the role of narration-based techniques like curiosity-triggering events and surprises that are included in games in learning and motivation. We focus on the learning of proportional reasoning, an important component of mathematical skills, with secondary prevocational students (12–15 year). Based on the information gap theory of Loewenstein and the cognitive conflict notion of Berlyne we argue that curiosity-triggering events and surprises can have a positive effect on learning. Inserting these events in the game *Zeldenrust* did indeed show positive learning effects, though the size of effect depends on the preexisting (meta)cognitive abilities of the students.

Keywords Serious games • Mathematics • Curiosity • Surprise • Learning • Motivation

6.1 Introduction

The question raised in this chapter is how we can stimulate players to engage in relevant cognitive processes that foster learning without jeopardizing the motivational appeal of the game by making use of narration-based techniques. From film and story understanding literature it is well known that stories can have an engaging influence on readers (Brewer & Lichtenstein, 1982). Stories also facilitate understanding and memory of the sequence of events that are part of the event structure that forms the basis of a story (Kintsch, 1980). Less clear from the same literature are the effects of certain techniques or directed manipulations of story structures that maximize the effects on emotion and learning. One exception here is the influence of techniques such as curiosity and surprise (Brewer & Lichtenstein, 1982; Hoeken & van Vliet, 2000; Kintsch, 1980). By starting a story with its outcome, readers become curious about how this event came about, leading to more attention

H. van Oostendorp (✉) • P. Wouters
Utrecht University, Utrecht 3508TB, The Netherlands
e-mail: h.vanOostendorp@uu.nl

P. Wouters, H. van Oostendorp (eds.), *Instructional Techniques to Facilitate Learning and Motivation of Serious Games*, Advances in Game-Based Learning, DOI 10.1007/978-3-319-39298-1_6

for the story. Surprise is evoked by an unexpected event. It can be created by inserting in a story an event that does not follow the normal flow of events nor is directly compatible with it. It stimulates the reader to wonder why this event happened, leading to enhanced and focused cognitive processing (Kim, 1999).

Also on the domain of serious games the role of stories and strong story lines in Game-Based Learning (GBL) is emphasized (Barab, Gresalvi, & Ingram-Goble, 2010; Dickey, 2006, 2011), though empirical support is still scarce (Adams, Mayer, MacNamara, Koenig, & Wainess, 2012; Pilegard & Mayer, 2016). Also the role of curiosity and surprise is neglected, though they represent on the domain serious games in our opinion promising techniques. To start with curiosity, it can be regarded as a motivator for active cognitive explorative behavior (cf. Berlyne, 1960; Litman, 2005; Loewenstein, 1994) and active exploration is a key aspect of contemporary computer games (Dickey, 2011). Next, surprise is generally conceived as a disruption of an active expectation. The situation described does not correspond to expectations of the reader (or player). Common to curiosity and surprise is that they both involve the experience of a cognitive discrepancy or conflict in the mental representation that the player is building up. That is, a cognitive conflict in the sense that events are introduced that can only be understood after extra cognitive processing which is needed to make the mental representation after all still coherent (Maguire, Maguire, & Keane, 2011).

In this chapter, we will first elaborate the concept of cognitive conflict and information gap that readers or players experience with surprises and curiosity-triggering events. Next we will present own work on the role of curiosity with regard to learning in serious games, and after that, discuss own work on the role of surprise in serious games. In our work on curiosity and surprise, we have followed a 'value-added' approach (Mayer 2011, 2016). In this research approach, we compare the learning outcomes of students who play a base version of a game (control group) with those who play the same game, but with one feature added, the curiosity or surprise-triggering events, respectively (treatment group). In the last section, we will draw conclusions and discuss implications.

6.2 Role of Curiosity and Surprise in Engaging Players in Game-Based Learning

In his review, Loewenstein (1994) proposes an information gap theory in which *curiosity* is supposed to arise when attention becomes focused on a gap in one's knowledge. Such an information gap produces the feeling of deprivation labeled curiosity. The curious individual is motivated to obtain the missing information in order to reduce the gap and to eliminate the feeling of deprivation.

This interpretation of an information gap is also related to Berlyne's concept of a cognitive conflict (Berlyne, 1960). This construct encompasses 'collative' variables such as complexity, novelty, and surprisingness. The presence of these stimulus characteristics (curiosity-triggering events) would arouse cognitive conflict and stimulate curiosity. In this case, an information gap occurs when stimuli, for instance, text fragments, present contradictory or incongruent information. For example, in the game a learner is told that a presented problem can be solved but the game environment appears to offer no opportunities to solve the problem. This interpretation of an information gap can also be regarded as a cognitive conflict, namely, the conflict in the current mental representation of the learner between (1) the expectations of the learner (e.g., expectations based on the assurance that the problem can be solved) opposed to (2) the affordances in the learning environment to solve the problem.

The assumption, in line with Jirout and Klahr (2012), is that this information gap will motivate students to explore the environment and find relevant information for constructing and applying appropriate solution methods. More specifically, we assume that based on Loewenstein's (1994) and Berlyne's (1960) ideas that externally inducing an information gap can stimulate curiosity, raise arousal, and consequently enhance explorations in the game environment and in this way improve learning.

In another context, using a puzzle game involving the ability to plan like solving the Tower of Hanoi problem, we have shown that omitting particular information on the display of a computer screen versus explicitly showing it did not affect the efficiency of the game adversely but did improve learning, especially of the underlying rules of the game (Van Nimwegen, 2008; Van Nimwegen, van Oostendorp, & Tabachneck-Schijf, 2005). We concluded that presenting visual support resulted in passive cognitive behavior. On the other hand, those who were refrained from this information and thus experienced an information gap were prompted to proactive and plan-based problem-solving behavior, leading to more effective cognitive processing. We assumed that creating an information gap by omitting relevant information in this study indeed would lead to a cognitive conflict that triggered effective exploratory behavior, and consequently to better learning.

A second promising technique in the generation of manageable cognitive conflicts consists of introducing *surprises*. We define surprise here as a disruption of an active expectation under influence of surprise-triggering events. Surprise also involves an emotional reaction and serves a cognitive goal as it directs attention to explain why the surprising event occurred and can play a key role in learning (Foster & Keane, 2015; Howard-Jones & Demetriou, 2009; Ranganath & Rainer, 2003).

The experience of surprise arises when an observed event causes a previous coherent representation to break down, resulting in an urgent representational updating process (Itti & Baldi, 2009; Maguire et al., 2011). Studies investigating the comprehension of narratives stress the idea that surprise is linked to ease of integration of the surprising event into the mental representation that is being built in Working Memory. Along the same lines, Kintsch (1980) also assumes that surprising events have important effects on the cognitive reading process. When reading a story, readers build a mental representation of it. The occurrence of a surprise triggering event forces readers to reassess their representation of the story up till that point, because a surprising event is by definition not a logical sequel to the preceding events. They have to check their representation to see whether they missed something. As a result of this reassessment and coherence checking, encoding and subsequently learning of the preceding events improves. On the domain of narratives and text comprehension it has been shown that surprise has a beneficial effect on learning. Hoeken and van Vliet (2000) found that surprise improved text comprehension and appreciation more than other techniques such as events that aroused curiosity and suspense. Likewise, O'Brien and Myers (1985) confronted participants with a word that was either predictable or unpredictable from a preceding context and observed that the texts that preceded unpredictable words were better recalled.

We assume that the effect of surprise may also pertain to problem solving or learning cognitive skills as mathematics in serious games. Ideally, mental models enable students to recognize specific characteristics of a problem and how to solve that problem. Because our aim is to integrate the instructional technique (i.e., the introduction of a curiosity-triggering event or surprise) with the learning content (Habgood & Ainsworth, 2011), the curiosity-triggering events and surprises have to be focused on what has to be learned, i.e., the mental model. For this reason, the curiosity-triggering events or surprising events change some of the problem characteristics and the solution method previously applied is no longer applicable and the player has to reevaluate the situation and decide which problem characteristics are relevant and also which solution method is now most appropriate. We expect thus that also surprise has a positive effect on learning because it may stimulate relevant cognitive processes such as organizing and integrating information (Mayer, 2011; Moreno & Mayer, 2007; see Fig. 1 in Chap. 1) without compromising the motivational appeal of computer games.

6.3 Role of Curiosity in Game-Based Learning

In this section, we will discuss some recent empirical work we performed on curiosity and learning mathematics, such as the skill of proportional reasoning. The goal of the studies discussed here was specifically to study the role of curiosity and we will detail the way we manipulated curiosity. As indicated earlier the advantage of curiosity induced by an information gap is that individuals are cognitively active in an engaging way. Scholars have emphasized the potential of curiosity in GBL (Dickey, 2011; Malone, 1981; Wouters, van Oostendorp, Boonekamp, & van der Spek, 2011) but empirical research is scarce. For instance, Wouters et al. (2011) showed empirically that introducing narrative elements as foreshadowing creates curiosity; however, it did not yield learning. The game for which we used the well-known game *Re-mission* (Beale, Kato, Marin-Bowling, Guthrie, & Cole, 2007) contained as foreshadowing technique briefly showing events in advance that occur later in the game. This foreshadowing technique can be regarded as an example of an information gap: some information is shown, but it is not sufficient to fully understand what is happening. Consequently, the attention of the players is drawn and they will be motivated to find the remaining information as soon as an opportunity arises. Compared to a control condition on a curiosity questionnaire the experimental condition showed higher curiosity. However on a (limited) factual recall test, there was no significant difference though the experimental condition showed somewhat better performance.

In next studies, the aim was to investigate more systematically whether the use of curiosity in a GBL environment enhances learning for proportional reasoning. Our operationalization of curiosity was based on Loewenstein's information gap theory (1994) as introduced earlier.

6.3.1 Game

The game involved proportional reasoning. This topic was chosen because it is a relevant and well-defined domain and existing methods for proportional reasoning are often ineffective (Rick, Bejan, Roche, & Weinberger, 2012). Several types of problems can be distinguished. For instance, in missing value problems one value in one of two proportions is missing and learners have to find this "missing value" in order to ensure that both proportions are equal (for a more extensive description see Vandercruysse et al., 2014, and Chap. 2 this volume). In the 2D game (Flash/Actionscript) called *Zeldenrust*, students have a summer job in a hotel (see http://www.projects.science.uu.nl/mathgame/zeldenrust/index.html for a demo). By doing different tasks the students can earn money that they can use to select a holiday destination during the game: the more money they earn, the further they can travel. During the game, the player is accompanied by the manager, a nonplaying character, who provides information about the task and gives feedback regarding the performance on the task. The game comprises a base game and several subgames. The base game provides the structure from which the subgames can be started that cover specific problem types in the domain of proportional reasoning. The tasks are directly related to proportional reasoning (e.g., mixing two drinks to make a cocktail according to a particular ratio directly involves proportional reasoning skills). In addition, mental operations with respect to proportional reasoning are connected with the game mechanics (e.g., in order to get the correct amount of bottles in the refrigerator the player has to drag the correct number of bottles in the refrigerator).

In the *control* condition, all assignments were presented in an identical way and all information required to perform the assignment was available. In the *curiosity* condition, the operationalization of curiosity involved two phases. First, the student was told that a strange situation had occurred but that the current problem could still be solved (Fig. 6.1a). In this way, we created an expectation in the student that was not immediately compatible with the situation in the game. Second, the student was confronted with a game environment in which it was not immediately clear how the assignment could be solved. Taken together, we regard this as a cognitive conflict, namely, the conflict between the expectations of the learner and the affordances in the learning environment. They have to explore the contents in the crates and decide how they can solve the problem the best. For example, the blackboard in Fig. 6.1b makes clear that four bottles of Cola have to be moved into the refrigerator; however, there are not directly available four bottles of Cola. The learner can hover

Fig. 6.1 The implementation of curiosity (**a**) depicts the initial situation, (**b**) shows the content when hovering over the crate with the mouse, and (**c**) shows the situation when the crate is unpacked

the crates to find out and reveal their content. The left crate in Fig. 6.1b contains three smaller packages with 4, 6, and 8 bottles. By exploring the different crates, the learner can decide which crate contains the packages that can best be used to solve the problem. With a mouse click the large crates are unpacked and the smaller packages become available, one of them with the right amount of bottles (Fig. 6.1c). Our assumption was that students had to explore the game environment and find and evaluate the objects (crates/bottles) that would enable them to implement the solution that they had conceived.

Before and after playing the game, a proportional reasoning skill test was administered in two groups of prevocational students. One group received the experimental version of the game (with the curiosity-triggering events) and the second group played the control version of the game (without these curiosity-triggering events).

6.3.2 Outcomes of Studies on Curiosity

The results of a first study with *Zeldenrust* showed that playing the game had a learning effect, that is, in both conditions there were significant gains in proportional reasoning skill; however, the curiosity condition did not advance more in proportional reasoning skill than the control group (Wouters et al., 2015a, 2015b). In a second study with improvements on instruction and interface design, game play did not yield learning, though in both studies performance on the game assignments contributed strongly to offline posttest performance (based upon multiple linear regression analyses). However, most important is that we in both experiments failed to find a beneficial effect of the curiosity-triggering events compared to a control condition that played the game without curiosity-triggering events embedded. The curiosity condition was not any better than the control condition.

As explained, based on Loewenstein's (1994) and Berlyne's (1960) ideas, we hoped that these *situational*, i.e., externally defined, determinants would induce curiosity. The game environment however had a strong repetitive character which made it perhaps difficult to maintain a curiosity effect. Our implementation of curiosity depended on a conflict or incongruity between what players was told and what they saw. Some remarks can be made regarding this implementation. Can an incongruity that is materialized in two different modalities (verbal and visual) evoke the intended cognitive conflict? It was maybe difficult for some students to make a connection between what was told in the verbal modality and what was shown in the visual modality. This may also explain the confusion that some students experienced when they were confronted with the curiosity-triggering events. It is worth to investigate the impact of the curiosity-triggering events when they occur in only one modality. We don't know exactly if players experienced a cognitive conflict or that they were just confused. For this reason, an obvious next step might be to understand what players think or experience when the curiosity-triggering events occur. Interesting methods in this respect are the use of think-aloud protocols and/or eye tracking.

In a third experiment using the same *Zeldenrust* game we manipulated the knowledge gap more directly by varying in the game the difficulty level of tasks compared to the current level of the player (De Wildt, 2015). The idea is that a higher difficulty level should make the knowledge gap larger. The basic idea of Loewenstein (1994) is that a difference between the current knowledge level and the knowledge needed to solve a particular problem may evoke curiosity and the desire to close the gap (when it is not too big or too small) which leads to extra attention and learning. Players were presented game tasks of a higher difficulty level (large knowledge gap) compared to the current skill level of the participant or—in the other condition—of the same difficulty level (small or no knowledge gap). We assume that a large knowledge gap leads to a clearer and more salient cognitive conflict than a small knowledge gap. So a situation representing a larger knowledge gap (bigger discrepancy in knowledge) makes the cognitive conflict more clear and more salient leading to a higher state of curiosity and an increase of learning.

We did indeed find a marginal significant positive effect of knowledge gap ($p < .08$) on learning as reflected by the performance on the proportional reasoning skill test. So introducing a knowledge gap can increase learning gain. We have the impression that the results can be enhanced further when more subtle measurements and adaptations of skill level are made. In the current study, we determined it beforehand and the study used only a small number (3) of difficulty levels. When adaptation occurs more smoothly and continuously, bigger positive effects may be expected, as for instance in the study of Van Oostendorp, van der Spek, and Linssen (2014). They showed a large positive effect of dynamic adaptation in terms of efficiency of players. The game used here was *Code Red: Triage*, a game focused on training a triage procedure for medical first responders. Compared to a control condition with no adaptation, an online, adaptive version of the game was (about 30 %) more efficient and lead to higher learning gains per instructional case.

Summarizing these studies to the role of curiosity, positive effects on learning can be found but the effects are subtle. They depend, for instance, on the clarity or saliency of the curiosity-triggering event and the knowledge level of the player.

6.4 The Role of Surprise in Game-Based Learning

In this section, we discuss studies in which we investigated the different dimensions of surprise in different games and domains.

Empirical research has demonstrated that indeed surprise can have specific effects on brain activity, also in a serious game context. Georgiadis, van Oostendorp, and van der Pal (2015) studied specific effects of surprise on brain activity as measured by EEG. In this study, a game was constructed in which the player acted as an undercover agent who had to perform a series of actions in order to save commercial supplies on an island from terrorists. Surprises consisted of inserting events that were unexpected compared to preceding events, like a sudden fire or explosion in a car. In a control version the surprises were left out. The results showed that surprises did lead to a more wakeful state indicated by lower Delta brainwaves (Hammond, 2006). Furthermore, experiencing surprises did lead to better in-game performance and better handling of later surprises by being more relaxed and conscious, as indicated by lower Alpha brainwaves (Benca et al., 1999), compared to a control version of the game without surprises. The last result indicates that training of surprises can have practical positive effects.

In the context of learning a medical procedure with a serious game called *Code Red: Triage*, Van der Spek, van Oostendorp, and Meyer (2013) demonstrated that surprise yielded superior knowledge structures, indicating that surprising events foster deep learning. They assumed that also in games players construct a mental model based on the story line, the events, and the underlying rules of the game. During understanding a story, readers construct a situation model in which dimensions such as the protagonist, time, space, causality, and intentionality are monitored and connected (Zwaan, Langston, & Graesser, 1995). When there are gaps constructing a connection takes more effort and time. Likewise, in computer games players construct a mental model and/or situation model based on the story line, the events, and the underlying rules of the game (Van der Spek et al., 2013). The situation or mental model makes new events plausible (although such events may cause adaptations in the model) and provides the starting point for expectations of the reader or player. A surprise, on the other hand, is unexpected and does not follow from the situation/mental model in a standard way. Readers/players will wonder what they have missed and start to reevaluate preceding events and infer events that make the surprising event understandable. In this process, the mental model will be activated, retrieved, and updated, thereby enhancing learning (Van der Spek, 2011; Van der Spek et al., 2013). As mentioned before we assume that this mechanism is also applicable to problem solving.

In two studies we investigated the impact of surprise on learning proportional reasoning and how this impact is moderated by the expectancy of the student (in the second study). We used the same GBL environment *Zeldenrust* as mentioned earlier.

In the first study, a group of prevocational students coming from different educational levels playing the game with surprises occurring during the game was compared with a group without these surprises (control group). We expected that the group with surprises would learn more than the control group.

6.4.1 Game

The control condition was the same as the control condition in the curiosity experiments. The *surprise* condition comprised a nonplaying niece character in the introduction animation telling that she sometimes will make it difficult to carry out the task. When the surprise occurred the niece character popped up and told that she had changed something. This change involved a sudden change of specific characteristics of the task whereby the solution method of the player doesn't apply anymore and the player has to reconsider the original solution method. The surprise is here thus a sudden change of some characteristics of a state in the situation. Figure 6.2 gives an example of the occurrence of a surprise. Figure 6.2a depicts the starting situation. The player can solve the problem by looking at the ratio "within": the number of Fanta in the refrigerator is twice as much as the number of Fanta in the desired proportion (12 Fanta) since $12*2=24$, so the number of Cola also has to be doubled ($9*2=18$ Cola). When the player is implementing the solution, the surprise occurs consisting of the nonplaying character suddenly changing the situation (Fig. 6.2b). When the niece character has disappeared the characteristics of the task are modified (Fig. 6.2c); that is, the desired proportion is now 5 Cola per 10 Fanta. The ratio "within" is not applicable anymore and the player can better use a method based on the ratio "between" (the desired proportion is 5 Cola/10 Fanta, so the number of Cola in the refrigerator should also be half the number of Fanta, 12/24). So the surprise does not simply concern some numbers. It also urges the player to suddenly change the solution method and replace that for a new one. In total the players received 8 surprises.

6.4.2 Outcomes of Studies on Surprise

The results of the first study indicated that surprise was beneficial for higher level students, while the main effect of surprise versus no surprise was not significant. For this reason, we repeated the study with only higher educational level students.

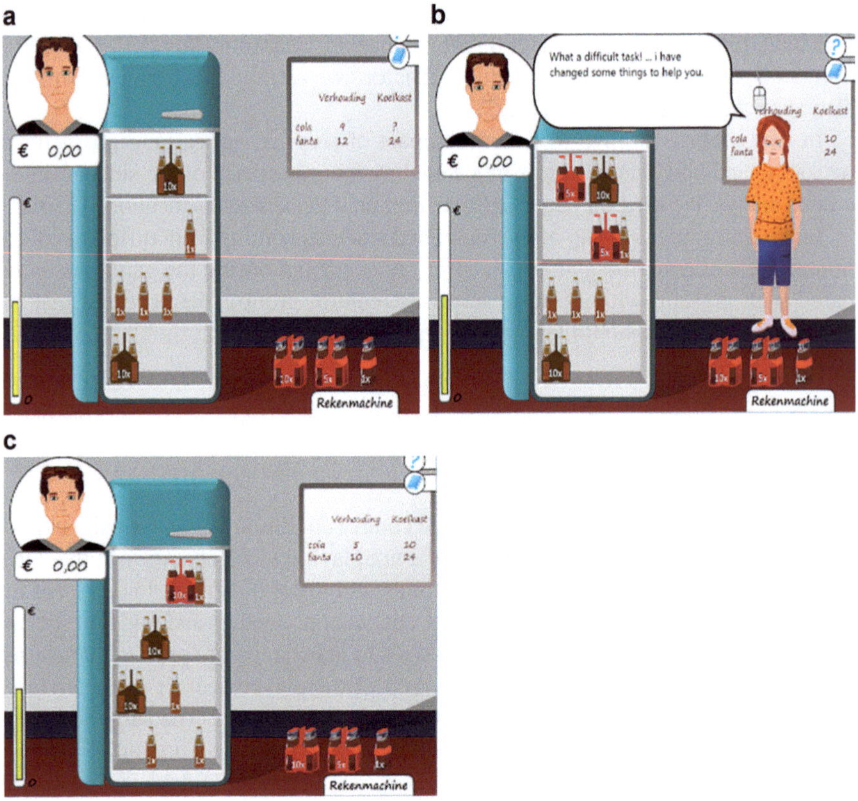

Fig. 6.2 (**a**) Starting situation in a task with a surprise, (**b**) notification of the surprising event, and (**c**) modification of task characteristics in the game *Zeldenrust*

We did find in this second study a significant positive effect of surprise on the posttest of reasoning skill when we included preexisting proportional reasoning skill as factor (Wouters et al. 2015a, 2015b).

Summarizing the results to the role of surprise shows that positive effects of surprise can be found, though the effect depends on the (meta)cognitive level of the students.

6.5 Conclusions

Overviewing the outcomes of our studies on curiosity and surprise with the game *Zeldenrust*, we can conclude that curiosity can have a moderate positive effect on learning; however, the effect depends on the clarity of the curiosity-triggering event and the knowledge level of the player. Surprise shows a positive effect but also here the effect depends on preexisting (meta)cognitive abilities of the student.

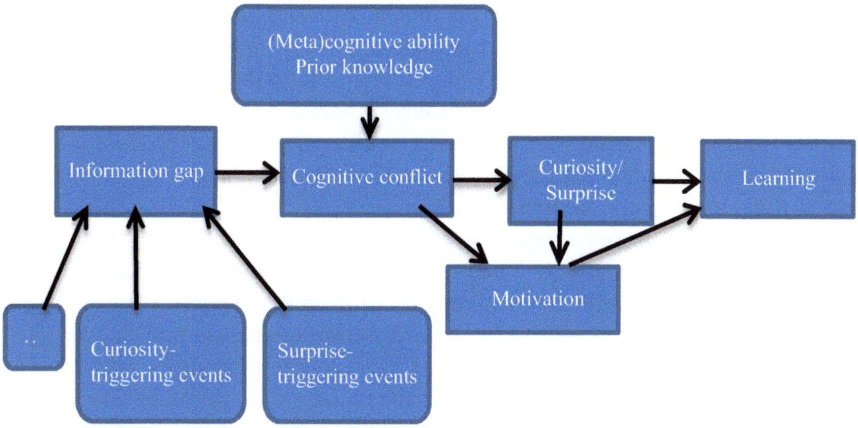

Fig. 6.3 Relation between narration-based techniques as curiosity and surprise-triggering events, and motivation and learning

When we compare the effects of curiosity-triggering events with the effects of surprises, it seems that we do find more easily positive learning effects of surprises while the effects of curiosity are weaker or not present at all. One reason could be that it is more difficult to trigger effectively curiosity with the inserted events; there is still much freedom for players to use them or not or to decide in what direction the inference processing in order to solve the problem should go. The surprises seem to be more constraining, and because of that, perhaps more effective. It is with a surprise immediately clear what the information gap is that has to be resolved, and thus what the cognitive conflict is. With a curiosity-triggering event, it depends on many factors (e.g., their saliency or clarity) whether the curiosity-triggering event leads to explore the information space in the right direction—and indeed to exploration at all—leading to a less well-defined information gap, and consequently, less clear cognitive conflict. See Fig. 6.3 in which we have depicted the assumed relationship between narration-based techniques like curiosity and surprise, and learning.

The results we found imply that instructional techniques such as curiosity and surprise should be applied with care. An important precondition for the occurrence of effective curiosity and surprise seems to be that players have sufficient cognitive flexibility and metacognitive abilities and prior knowledge to orientate on the task, to reevaluate the results at the moment when the surprise or curiosity-triggering event occurs and to reflect on the performed actions. See also Fig. 6.3. Students with sufficient (meta)cognitive abilities seem to be able to handle surprises and curiosity-triggering events in complex learning environments such as computer games, students who lack these competencies can be overwhelmed by the additional cognitive demands that are introduced by these techniques. However, more research to the relation between (meta)cognitive abilities, and curiosity and surprise is required to investigate the robustness of the surprise and curiosity effects and the underlying cognitive mechanisms.

We have discussed in Sect. 6.2, the role of curiosity and surprise in engaging players in game-based learning, and as indicated in Fig. 6.3 we assume that narrative techniques as including curiosity and surprise-triggering events can also have a positive influence on motivation, and because of that, also on learning. In our studies we focused on learning so we cannot confirm the indirect positive effect of curiosity and surprise on learning. Future studies should look into the role of motivation triggered by these techniques.

Curiosity and surprise require as indicated earlier cognitive flexibility and (meta) cognitive abilities because they imply a deviation from what students expect. Students who do not have an adequate level of cognitive flexibility and/or (meta) cognitive abilities may benefit from additional instructional support that will help them to understand the problem and possible solution methods. In this way, the consequences of curiosity and surprise for the problem may become clearer. In our own research on surprise, we expected students to select an appropriate method for a given problem type but we found that some students always used the same method regardless of the problem type. This attenuates the effect of surprise because the purpose of surprise—to consider another method when the problem type suddenly changes—is beyond their cognitive ability. To support these students, the surprise intervention can be preceded by exercises that will help them to select an appropriate method for a problem. One could think of exercises that help them to automatize part tasks such as multiplication tables so that they can more easily identify "intern" or "extern" ratios and/or worked examples in which strategies for specific types of problems are modeled.

Two other lines of research can be directly relevant to our research on narration-based techniques such as surprise and curiosity. The first one concerns the role of (meta)cognitive abilities. There is some evidence that metacognitive skills in mathematics improve with small differences in age (Van der Stel et al., 2010). The students in the second study on surprise had a mean age of 13.9 years (second year class) and the metacognitive skills of some may have been insufficiently developed. Another point is that the students come from the least advanced of three Dutch prevocational tracks in which students are prepared for intermediate vocational education. It would be interesting to replicate our studies on curiosity and surprise with older students in the same educational level (third or fourth year class) or students from a higher educational track.

A second research avenue pertains to the characteristics of the game. The game *Zeldenrust* we used as test bed to investigate the usefulness of narration-based techniques such as curiosity and surprise has unfortunately a repetitive character; that is, students engage in the same type of tasks which require similar actions. It is not unlikely that students finally will expect that, for instance, the surprising niece character will reappear and modify the nature of the task. In that case they may even anticipate these events and thus undermine the potential effect of surprise. The same applies to the curiosity manipulation. If that is the case more variation in surprise or curiosity can perhaps further increase their effectiveness.

It may seem that the introduction of curiosity or surprise adds more difficulty to the problems presented. Two comments to this suggestion can be made. First, making a problem in first instance somewhat—not too much because the gap should not be too large—e.g., by omitting particular information, can improve learning, particularly of the underlying rules of the problem (Van Nimwegen, van Oostendorp, & Tabachneck-Schijf, 2005). Second, we want to point out that our results showed surprise did have a positive effect on learning, particularly of students with sufficient (meta)cognitive abilities.

Acknowledgement This research is funded by the Netherlands Organization for Scientific Research (project number 411-00-003).

References

Adams, D. M., Mayer, R. E., MacNamara, A., Koenig, A., & Wainess, R. (2012). Narrative games for learning: Testing the discovery and narrative hypotheses. *Journal of Educational Psychology, 104*(1), 235–249.

Barab, S. A., Gresalvi, M., & Ingram-Goble, A. (2010). Transformational play: Using games to position person, content, and context. *Educational Researcher, 39*(7), 525–536.

Beale, I. L., Kato, P. M., Marin-Bowling, V. M., Guthrie, N., & Cole, S. W. (2007). Improvement in cancer-related knowledge following use of a psychoeducational video game for adolescents and young adults with cancer. *Journal of Adolescent Health, 41*, 263–270.

Benca, R. M., Obermeyer, W. H., Larson, C. L., Yun, B., Dolski, I., Kleist, K. D., et al. (1999). EEG alpha power and alpha power asymmetry in sleep and wakefulness. *Psychophysiology, 36*, 430–436.

Berlyne, D. E. (1960). *Conflict, arousal and curiosity*. New York: McGraw-Hill.

Brewer, W. F., & Lichtenstein, E. H. (1982). Stories are to entertain: A structural-affect theory of stories. *Journal of Pragmatics, 6*, 473–483.

De Wildt, R. (2015). *An analysis of the curiosity stimulating characteristics of serious games and their effect on learning and motivation*. Thesis Information Science, Dept. of Information and Computing Sciences, Utrecht University, Utrecht, The Netherlands.

Dickey, M. D. (2006). Game design narrative for learning: Appropriating adventure game design narrative devices and techniques for the design of interactive learning environments. *Educational Technology Research and Development, 54*(3), 245–263.

Dickey, M. D. (2011). Murder on Grimm Isle: The impact of game narrative design in an educational game-based learning environment. *British Journal of Educational Technology, 42*, 456–469.

Foster, M. I., & Keane, M. T. (2015). Predicting surprise judgments from explanation graphs. In *International Conference on Cognitive Modeling (ICCM)*, Groningen University, Groningen, The Netherlands.

Georgiadis, K., van Oostendorp, H., & van der Pal, J. (2015). EEG assessment of surprise effects in serious games. In *GALA2015 Conference,* Rome, Italy.

Habgood, M. P. J., & Ainsworth, S. E. (2011). Motivating children to learn effectively: Exploring the value of intrinsic integration in educational games. *Journal of the Learning Sciences, 20*, 169–206.

Hammond, D. C. (2006). What is neurofeedback? *Journal of Neurotherapy, 10*(4), 25.

Hoeken, H., & van Vliet, M. (2000). Suspense, curiosity, and surprise: How discourse structure influences the affective and cognitive processing of a story. *Poetics, 26*, 277–286.

Howard-Jones, P., & Demetriou, S. (2009). Uncertainty and engagement with learning games. *Instructional Science, 37*(6), 519–536.

Itti, L., & Baldi, P. (2009). Bayesian surprise attracts human attention. *Vision Research, 49*, 1295–1306.

Jirout, J., & Klahr, D. (2012). Children's scientific curiosity: In search of an operational definition of an elusive concept. *Developmental Review, 32*(2), 125–160.

Kim, S. (1999). Causal bridging inferences: A cause of story interestingness. *British Journal of Psychology, 3*, 430–454.

Kintsch, W. (1980). Learning from text, levels of comprehension, or: Why anyone would read a story anyway. *Poetics, 9*, 87–98.

Litman, J. A. (2005). Curiosity and the pleasures of learning: Wanting and linking new information. *Cognition and Emotion, 19*, 793–814.

Loewenstein, G. (1994). The psychology of curiosity: A review and reinterpretation. *Psychological Bulletin, 116*, 75–98.

Maguire, R., Maguire, P., & Keane, M. T. (2011). Making sense of surprise: An investigation of the factors influencing surprise judgments. *Journal of Experimental Psychology. Learning, Memory, and Cognition, 37*(1), 176–186.

Malone, T. (1981). Toward a theory of intrinsically motivating instruction. *Cognitive Science, 4*, 333–369.

Mayer, R. E. (2011). Multimedia learning and games. In S. Tobias & J. D. Fletcher (Eds.), *Computer games and instruction* (pp. 281–305). Greenwich, CT: Information Age.

Mayer, R. E. (2016). What should be the role of computer games in education? *Policy Insights from the Behavioral and Brain Sciences 3*(1), 20–26.

Moreno, R., & Mayer, R. E. (2005). Role of guidance, reflection, and interactivity in an agent-based multimedia game. *Journal of Educational Psychology, 97*, 117–128.

Moreno, R., & Mayer, R. E. (2007). Interactive multimodal learning environments. *Educational Psychology Review, 19*(3), 309–326.

O'Brien, E. J., & Myers, J. L. (1985). When comprehension difficulty improves memory for text. *Journal of Experimental Psychology. Learning, Memory, and Cognition, 11*, 12–21.

Pilegard, C., & Mayer, R.E. (2016). Improving academic learning from computer-based narrative games. *Contemporary Educational Psychology, 44*, 12–20. doi:10.1016/j.cedpsych.2015.12.002

Ranganath, C., & Rainer, G. (2003). Neural mechanisms for detecting and remembering novel events. *Nature Reviews. Neuroscience, 4*, 193–203.

Rick, J., Bejan, A., Roche, C., & Weinberger, A. (2012). Proportion: Learning proportional reasoning together. In A. Ravenscroft, S. Lindstaedt, C. D. Kloos, & D. Hernández-Leo (Eds.), *Lecture notes in computer science* (21st century learning for 21st century skills, Vol. 7563, pp. 513–518). Berlin, Germany: Springer.

Van der Spek, E. D. (2011). *Experiments in serious game design. A cognitive approach.* Doctoral dissertation, Utrecht University, Utrecht, The Netherlands.

Van der Spek, E. D., van Oostendorp, H., & Meyer, J.-J. C. (2013). Introducing surprising events can stimulate deep learning in a serious game. *British Journal of Educational Technology, 44*, 156–169.

Van der Stel, M., Veenman, M. V., Deelen, K., & Haenen, J. (2010). The increasing role of metacognitive skills in math: A cross-sectional study from a developmental perspective. *ZDM The International Journal on Mathematics Education, 42*(2), 219–229.

Van Nimwegen, C. (2008). *The paradox of the guided user: Assistance can be counter-effective.* Doctoral dissertation, Utrecht University, Utrecht, The Netherlands.

Van Nimwegen, C., van Oostendorp, H., & Tabachneck-Schijf, H. J. M. (2005). The role of interface style in planning during problem solving. In B. Bara, L. Barsalou, & M. Bucciarelli (Eds.), *Proceedings of the 27th Annual Cognitive Science Conference* (pp. 2271–2276). Mahwah, NJ: Lawrence Erlbaum.

Van Oostendorp, H., van der Spek, E. D., & Linssen, J. (2014). Adapting the complexity level of a serious game to the proficiency of players. *European Alliance for Innovation Endorsed Transactions on Serious Games, 1*(2), 1–8.

Vandercruysse, S., ter Vrugte, J., de Jong, T., Wouters, P., van Oostendorp, H., & Elen, J. (2014). 'Zeldenrust': A mathematical game-based learning environment for vocational students. In *Proceedings of Research on Domain-Specific Serious Games: State-of-the-Art and Prospects Conference,* University of Leuven, Belgium.

Wouters, P., van Nimwegen, C., van Oostendorp, H., & van der Spek, E. D. (2013). A meta-analysis of the cognitive and motivational effects of serious games. *Journal of Educational Psychology, 105*, 249–265.

Wouters, P., & van Oostendorp, H. (2013). A meta-analytic review of the role of instructional support in game-based learning. *Computers & Education, 60*(1), 412–425.

Wouters, P., van Oostendorp, H., Boonekamp, R., & van der Spek, E. D. (2011). The role of game discourse analysis and curiosity in creating engaging and effective serious games by implementing a back story and foreshadowing. *Interacting with Computers, 23*, 329–336.

Wouters, P., van Oostendorp, H., ter Vrugte, J., Vandercruysse, S., de Jong, T., & Elen, J. (2015a). The role of curiosity-triggering events in game-based learning for mathematics. In J. Torbeyns, E. Lehtinen, & J. Elen (Eds.), *Describing and studying domain-specific serious games* (pp. 191–208). New York, NY: Springer.

Wouters, P., van Oostendorp, H., ter Vrugte, J., Vandercruysse, S., de Jong, T., & Elen, J. (2015b). The role of surprise in game-based learning for mathematics. In *GALA2015 Conference*, Rome, Italy.

Zwaan, R. A., Langston, M. C., & Graesser, A. C. (1995). The construction of situation models in narrative comprehension: An event-indexing model. *Psychological Science, 6*(5), 292–297.

Chapter 7
Designing Effective Feedback Messages in Serious Games and Simulations: A Research Review

Cheryl I. Johnson, Shannon K.T. Bailey, and Wendi L. Van Buskirk

Abstract Taking a value-added approach, we examined the impact of feedback on learning outcomes and performance in serious games and simulations. Although feedback has been demonstrated to be beneficial in traditional learning environments, we explore how feedback has been implemented in game- and simulation-based learning environments. In this review, we discuss critical characteristics that affect the efficacy of feedback, including the content of feedback messages, the modality in which feedback is presented, the timing of feedback presentation, and learner characteristics. General guidelines based on the research evidence are provided, and the theoretical implications are discussed in the context of the cognitive theory of multimedia learning (CTML; The Cambridge handbook of multimedia learning, Mayer, 2014b).

Keywords Feedback • Modality of feedback • Timing of feedback • Content of feedback • Adaptation

7.1 Introduction

Over the last decade, the promise of games and simulations for education and training has been touted by researchers and practitioners alike. Many boast about the motivational potential of games compared to traditional classroom instruction and argue that playing games is more intrinsically motivating. It follows that players will devote more cognitive effort while they are playing and are more likely to play

C.I. Johnson (✉) • W.L. Van Buskirk
Naval Air Warfare Center Training Systems Division,
Code AIR-4.6.5.1, 12211 Science Drive, Orlando, FL 32826, USA
e-mail: cheryl.i.johnson@navy.mil

S.K.T. Bailey
Department of Psychology, University of Central Florida,
4000 Central Florida Blvd, Orlando, FL 32816, USA

P. Wouters, H. van Oostendorp (eds.), *Instructional Techniques to Facilitate Learning and Motivation of Serious Games*, Advances in Game-Based Learning, DOI 10.1007/978-3-319-39298-1_7

again, leading to higher learning outcomes (Garris, Ahlers, & Driskell, 2002; Gee, 2007; Prensky, 2001). There have been several meta-analyses examining whether or not games and simulations truly are more effective for learning than conventional methods of instruction (e.g., Hays, 2005; Ke, 2009; O'Neil & Perez, 2008; Sitzmann, 2011; Vogel et al., 2006; Wouters, van Nimwegen, van Oostendorp, & van der Speck, 2013), and the findings have been mixed. However, several reviews suggest that games and simulations are more effective for learning when they include instructional support (e.g., Hays, 2005; Ke, 2009; Sitzmann, 2011; Wouters et al., 2013; Wouters & van Oostendorp, 2013). The focus of this chapter is to explore what we know about feedback, arguably one of the most important instructional features in serious games.

7.1.1 Defining Serious Games

A *serious game* is a game that is played with the intention to learn knowledge or a skill. Serious games are referred to by many names, including educational games, simulation games, game-based learning, and games for learning. The serious game literature is riddled with inconsistent definitions. There is no clear consensus in the extant literature on the definition of a game. Several definitions do converge on some of the defining characteristics of games that include: games are interactive (the player can make actions), rule based (events are bound within a set of rules), challenging (opportunities exist for success and for overcoming difficulties), responsive (the game responds to player actions), and goal oriented (the player works to achieve a knowable goal and can monitor progress along the way; Garris et al., 2002; Mayer, 2014a; Mayer & Johnson, 2010; Sitzmann, 2011; Vogel et al., 2006; Wouters et al., 2013).

Further complicating the issue is that the line between games and simulations is not always well-defined (Sitzmann, 2011; Vogel et al., 2006). *Simulations* can be defined as representations of a real-world system or process that allow the learner to test variables to see how the system will respond or to practice certain behaviors or skills that may be dangerous or cost-prohibitive to perform in real life (Merchant, Goetz, CiFuentes, Keeney-Kennicutt, & Davis, 2014; Sitzmann, 2011; Vogel et al., 2006; Wouters et al., 2013). This definition of a simulation shares several common attributes with that of serious games, including interactivity, challenge, and responsiveness. Following a similar logic, previous researchers have decided to combine serious games and simulations within their analyses (Garris et al., 2002; Sitzmann, 2011; Wouters et al., 2013), and we have chosen to do so here as well. Since it is not the goal of this chapter to resolve this debate, for simplicity's sake, we include game and simulation studies in our analysis and draw those distinctions when possible.

Not surprisingly, the serious game and simulation literature is vast and diverse. To address this diversity in the literature, Mayer and Johnson (2010) identified three different approaches to empirical games research that vary in terms of research goals. The goal of the *media comparisons* approach is to examine whether people learn better by playing games or learning from conventional instruction. The goal of the *cognitive consequences* approach is to investigate if playing a particular game causes players to improve one or more cognitive skills, such as spatial ability, perceptual attention, or motor skills. The goal of the *value-added* approach is to examine if adding an instructional feature to the game improves learning compared to a group who plays the same game but without the added instructional feature. This chapter will focus on research that takes a value-added approach to determine situations in which feedback improves student learning.

7.1.2 Defining Feedback

The purpose of feedback in serious games and simulations is to direct learners to improve performance, motivation, or learning outcomes by various methods of providing information to learners about the correctness of their responses (Shute, 2008). Providing feedback allows a learner to evaluate his or her progress and responses, identify knowledge gaps, and repair faulty knowledge (Johnson & Priest, 2014). From a broad review of the feedback literature focusing on serious games, simulations, and multimedia learning, we have identified four characteristics of feedback under which most feedback types fall. Methods of providing feedback vary by (1) type of feedback given (e.g., procedural or conceptual information, amount of elaboration), (2) the timing of feedback after a response (e.g., immediately following a question, at the end of a practice session), (3) the modality in which information is presented (e.g., spoken feedback, text-based feedback), and by (4) adapting to learner characteristics (e.g., prior knowledge, spatial ability). Feedback has been demonstrated to be an effective instructional technique to benefit learning (Azevedo & Bernard, 1995), and several meta-analyses have been conducted to determine when, how, and why it works (Bangert-Drowns, Kulik, Kulik, & Morgan, 1991; Hattie & Timperley, 2007; Kluger & DeNisi, 1996; Narciss & Huth, 2004; Shute, 2008). Although feedback has been studied mostly in more traditional educational settings, recent meta-analyses taking a value-added approach have found that adding instructional support, such as feedback, to a serious game improves learning outcomes (Ke, 2009; Mayer, 2014a; Wouters & van Oostendorp, 2013).

7.1.3 Present Chapter

The goal of this chapter is to address what is currently known about feedback in the serious games and simulations literature and draw conclusions about feedback strategies that can be useful for improving learning in the design and research of serious games. Throughout the chapter, we present empirical evidence from studies that take a value-added perspective to illustrate when certain feedback messages may be more effective than others. Although we originally set out to investigate the impact of feedback in serious games on learning *and* motivation, we found little empirical evidence on how providing feedback affects motivation. Therefore, since the literature on feedback and motivation is not yet mature enough to draw any strong conclusions, we chose to focus on the effects of different feedback strategies on performance and learning outcomes.

In this chapter, we outline the theoretical perspectives motivating the incorporation of feedback into games, highlight the core characteristics of feedback (content, modality, timing, and adaptation to individual differences), and provide practical implications of this research for future research and games design.

7.2 Theoretical Motivation

In order to understand how individuals learn from games—and how feedback can be effective in games—we must first consider how learning works. The cognitive theory of multimedia learning (CTML), as depicted in Fig. 7.1, provides a theoretical foundation for understanding how people learn (Mayer, 2014a, 2014b). The CTML has three main assumptions about a learner's cognitive system. First, learners possess separate channels for processing information: a visual channel and a verbal channel. The visual channel is used to process visual information; in the case of a serious game, learners would utilize the visual channel to view their avatar as they navigate the game environment. Likewise, the verbal channel is used to process verbal information, such as the spoken instructions from a pedagogical agent or teammate. The second assumption is that learners are limited in the amount of cognitive processing that occurs in any one channel at a time. That is, one or both channels can become overloaded if the learner's cognitive processing demands are too high. The third assumption is that learners are active participants during a learning episode and engage in the cognitive processes of *selecting* the relevant information from the serious game, *organizing* that information into a coherent mental representation, and *integrating* and updating this representation with new incoming information and with prior knowledge stored in long-term memory.

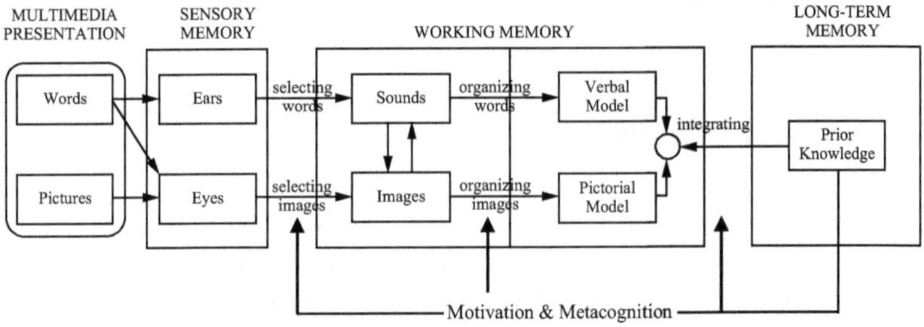

Fig. 7.1 The cognitive theory of multimedia learning (CTML). Adapted from Mayer (2014b)

Additionally, the CTML considers three demands on a learner's cognitive system during a learning episode. *Extraneous processing* is cognitive processing that occurs as a result of poor instructional or game design that does not contribute to the overall educational goal of the serious game or simulation. For example, a game that includes distracting sounds, unnecessary cut scenes, or complicated game controls and interfaces may create a situation of extraneous processing, since these features of the game are not productive toward the learning objectives of the game. *Essential processing* is cognitive processing that is required to mentally represent the information that is being learned in a learner's working memory that is the result of the material's complexity. Consider a learner playing a serious game that teaches how to troubleshoot electric circuits. The learner would need to represent mentally the elements of the circuit and their configuration in order to solve a troubleshooting question. For a novice, this situation is more complex and would likely require more essential processing than if the learner were playing a less complex game, such as a game to learn Spanish-English vocabulary words, which would require holding only two elements in working memory simultaneously. *Generative processing* is the cognitive processing directed at making sense of the essential information being presented to the learner in the serious game including reorganizing it and relating it to prior knowledge, which is dependent on the learner's level of motivation to play the game. If the learner is highly engaged in the serious game, the learner may put more effort into playing than if the game is not very engaging. It is important to reiterate that both essential and generative processing are productive toward the educational goal, while extraneous processing is not. Given that the learner's cognitive system has a limited capacity, extraneous, essential, and generative processing have an additive effect, such that if the extraneous processing demands overwhelm the learner's capacity, the learner will not have the cognitive resources available to engage in the more productive essential and generative processing. This situation can result in lower learning outcomes. Therefore, the goal for serious game designers is to reduce extraneous processing, manage essential processing, and foster generative processing in serious games to promote meaningful learning.

Mayer (2014b) recently expanded the CTML to include motivation and metacognition within the model (see also Moreno & Mayer, 2007). That is, while it is important to consider the types of cognitive processing that occur during learning, it is also important to acknowledge that one must have a desire to engage in the necessary cognitive processing to learn (i.e., motivation) and an understanding of how to manage one's cognitive processing appropriately (i.e., metacognition). Although motivation is also included in the definition of generative processing, the overall role motivation and metacognition play within the CTML model is "underdeveloped" (Mayer, 2014b, p. 66). In Fig. 7.1, an arrow points from long-term memory to the cognitive processes of selecting, organizing, and integrating, illustrating how motivation may affect these processes, and therefore how people learn. More research is needed to define the mechanism by which motivation may affect these processes and how feedback (and other instructional strategies) affects motivation.

Using CTML as a framework, one can consider how feedback could be useful for improving learning under certain conditions and for particular learners. At a high level, feedback serves to aid learners in the process of knowledge construction during a learning episode (Moreno & Mayer, 2005, 2007). In a *discovery-learning* environment, such as a serious game or simulation, the learner may freely explore the environment and receive minimal guidance on what to do and what to learn; in fact, in this free-play type of environment, it is possible that the learner may never come into contact with the to-be-learned material. Such a situation is likely to impose a high extraneous load on the learner, particularly on a novice learner, because he lacks knowledge of what is important to pay attention to within the game (Kirschner, Sweller, & Clark, 2006). Using a *guided* discovery-learning method, such as providing feedback in a serious game, learners can be pointed in the right direction and be provided an assessment of their performance. In addition, with feedback, a learner can also be provided with information about why a certain action or decision was correct or incorrect, and concepts, principles, and strategies can be explicitly explained, so that the learner is not left to make such a discovery on his own. That is, feedback can serve to reduce extraneous processing by guiding the learner to select the appropriate information in the learning environment (Kirschner et al., 2006; Mayer, 2003; Moreno, 2004) and may even serve as a way to help learners with organizing and integrating information (aiding in high level knowledge construction).

7.3 Related Work

7.3.1 Content of Feedback

Feedback content refers to the type and amount of information provided after a learner's response and varies considerably from merely providing a performance score (a form of outcome-based feedback) to providing an explanation of the correct answer (a form of process feedback). Similar to the issues found in the serious games literature, the feedback literature is also plagued with an inconsistent use of terminology. Therefore, in an effort to refine the numerous definitions of feedback content into usable guidelines for serious game designers and researchers, we consolidated and refined terms into a table of feedback content types, definitions, and examples (see Table 7.1). For each type of feedback, we have included alternative names used to refer to similar constructs. The types of feedback are grouped into outcome-based feedback and process-based feedback. *Outcome feedback* informs learners about their performance outcome or learning progress. *Process feedback* directs learners on the processes and/or strategies used to reach the correct answer or action in the game. Also, it is important to note that feedback content types may be combined and are not mutually exclusive. For example, feedback in a serious game may provide both outcome feedback and process feedback by informing the learner that his or her response is incorrect and explaining the steps needed to arrive at the correct answer. Although this may not represent an exhaustive list of

Table 7.1 Feedback content table

Feedback type		Definition	Example
Outcome	Knowledge of Result/Response (Verification, Corrective)	Feedback is given about whether the response was correct or incorrect	"Your answer was incorrect"
	Knowledge of Correct Result/ Response (Correct Response)	The correct answer is provided	"The correct answer was B"
	Answer until Correct (Try Again)	The learner must continue answering until he/ she answers correctly	The learner must infer that his/her response is incorrect until a correct response is provided and the learner is allowed to continue
	Percent Accuracy	Feedback is given to the learner about how well he/she is performing in a percentage of accuracy	"You were correct on 75 % of the questions"
	Error Flagging (Location of Mistakes)	Feedback shows where the errors in the response are without giving the correct response	"The first part of your answer is incorrect"
	Environmental	The actual relationship between response and outcome is modeled	When building an electric circuit, the bulb will not light up if the path is broken
	Velocity	The learner's current performance is compared to his/her performance on the previous trial	"Your performance declined from the previous scenario"
	Normative	The learner's ranking is compared to the standing of other learners	"You performed better than 80 % of other students"
Process (Elaborative, Explanatory)[a]	Informational	Feedback does not reveal the correct answer but gives some information on how to derive the correct answer or improve general understanding	"Remember to apply Ohm's law when solving for the required voltage"
	Topic Specific	Feedback provides more specific information about the target question or topic and leads the learner through the correct answer	"If you add a battery in serial, the flow rate increases"
	Response Specific	Feedback addresses why the correct answer is right and why the incorrect answer is wrong	"'A path that is not open' is not entirely correct. A bulb and a battery must be contained in the same closed path in order for a light bulb to light"
	Hints/Cues/Prompts (Worked-out example)	Prompts for the correct response or worked examples are provided to guide the learner to the answer, without explicitly stating the answer	"Here is a hint: A bulb is contained in what?"

Note: Table adapted from Mason and Bruning (2003), Shute (2008), and Van Buskirk (2011). These feedback types are not mutually exclusive. Feedback content may be combined. Types of feedback may also be labeled differently in feedback studies

[a]Process feedback may be conceptual or procedural

feedback content possibilities, it represents our approach to interpreting the variants of feedback used in the experiments discussed throughout this chapter. Below we describe the extant literature on feedback as it relates to serious games using the terminology of the original studies (and the new designation as represented in Table 7.1 where appropriate).

7.3.1.1 Comparing Explanatory Feedback to Corrective Feedback in Science Learning

Explanatory feedback (i.e., process feedback) occurs when a learner performs some action in a game or simulation and receives principle-based feedback describing the quality of his or her performance. Corrective feedback (i.e., outcome feedback), on the other hand, merely informs the learner whether the action he or she took was correct or not. Previous research has demonstrated that explanatory feedback is a very effective strategy to promote learning (Bangert-Drowns et al., 1991; Cameron & Dwyer, 2005; Mayer, 2014a; Shute, 2008).

According to the *feedback principle* in multimedia learning, novices learn more deeply with explanatory feedback than with corrective feedback (Johnson & Priest, 2014). In an exemplary study by Mayer and Johnson (2010), novice college students learned about electric circuits by playing the Circuit Game. The Circuit Game is an arcade-style game that consists of nine training levels, in which students answer questions about the rate of current of circuits in various configurations. A tenth level serves as an embedded near transfer test and students must determine, given two circuit diagrams, which light bulb burns the brightest. One group of students received only corrective feedback about their answers (i.e., corrective feedback group) with either a “ding” to denote a correct answer or a “buzz” to denote an incorrect answer (i.e., knowledge of correct response feedback group). A second group of students received corrective feedback in addition to explanatory feedback (i.e., explanatory feedback group) that showed the correct answer with an explanation of why that answer was correct (e.g., If you add a battery in serial, the flow rate increases; topic specific feedback). Figure 7.2 shows example screenshots from each condition. The authors found that those who received explanatory feedback during training performed better on the nine training levels and on the embedded transfer test than those who received only corrective feedback. They concluded that providing explanatory feedback guides the students to develop a deeper understanding of the material than merely providing corrective feedback.

Another example of the benefits of explanatory over corrective feedback is a set of studies by Moreno (2004). In two experiments, college students learned about botany by playing the Design-A-Plant simulation game. In this game, learners travel to a new planet with certain environmental conditions, and the learner is tasked to design a plant that would survive such conditions. Along the way, the learners are guided by an animated pedagogical agent, who provides feedback to learners on

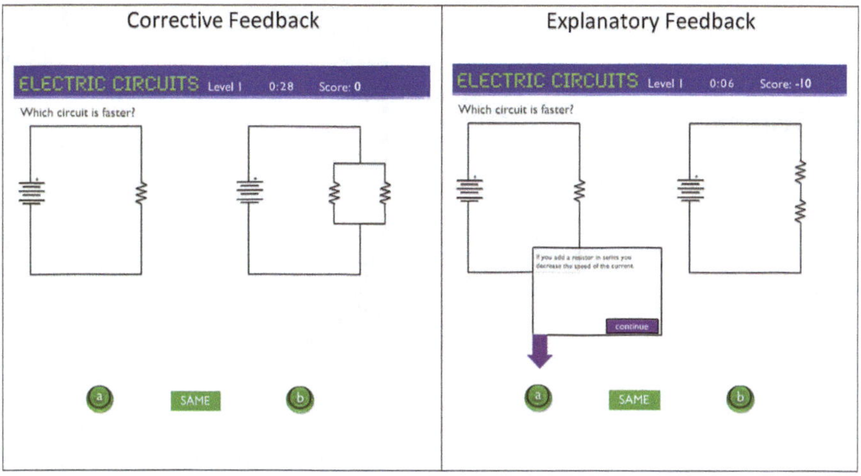

Fig. 7.2 Screenshots from the Circuit Game illustrating the corrective feedback condition (*left*) and the explanatory feedback condition (*right*). Adapted from Mayer and Johnson (2010)

their choices of roots, stems, and leaves that are most appropriate for the eight given environments learners travel to in the course of the game. Following the game, students completed a retention test, in which they had to recall the types of roots, stems, and leaves they learned about; and a transfer test, in which they designed plants for other nonstudied environmental conditions and had to determine, given a plant, the environment for which it was best suited. Students participated in one of two different feedback conditions. In the corrective feedback condition, the pedagogical agent informs the student whether his or her answer was correct or incorrect and then shows the student the correct answer (i.e., knowledge of correct response feedback). In the explanatory feedback condition, students received explanatory feedback from the pedagogical agent that explained why the student's choice of plant part was correct or incorrect (i.e., topic specific feedback). It should be noted that in both cases, the learners receive feedback about the correctness of their answer, but the explanatory feedback condition receives additional information about why the answer was correct or incorrect. Across both experiments, the explanatory feedback condition performed better than the corrective feedback condition on the transfer test, produced fewer incorrect responses during learning, and rated the game as more helpful. These results were replicated in a follow-up experiment conducted by Moreno and Mayer (2005, Experiment 1), and they also found that those who received explanatory feedback, as compared to those who received only corrective feedback, produced significantly more correct explanations for their answers on the transfer test and showed evidence for a reduction in the number of misconceptions. In summary, research evidence demonstrates that providing explanatory feedback that explains why a particular answer is correct promotes learning as compared to only providing corrective feedback.

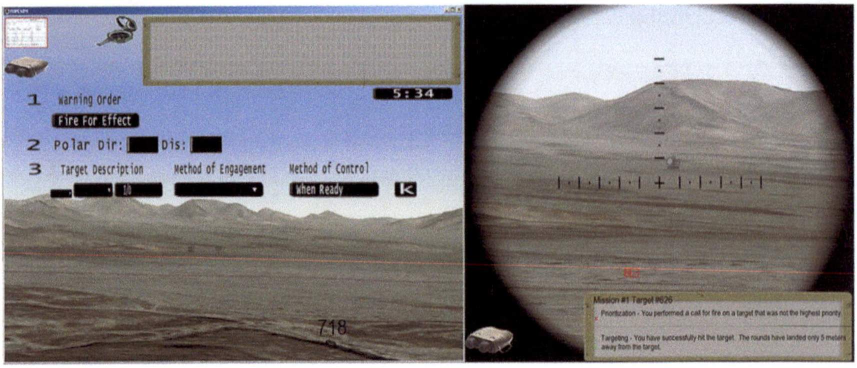

Fig. 7.3 Screenshots from FOPCSIM illustrating the call-for-fire sheet (*left*) used in Astwood et al. (2008), Landsberg, Van Buskirk, and Astwood (2010), and Van Buskirk (2011) and feedback (*right*; showing immediate, visual, outcome feedback for correct targeting and incorrect prioritization) used in Van Buskirk (2011)

7.3.1.2 Comparing Process and Different Types of Outcome Feedback in a Military Decision-Making Simulation

In a study by Astwood, Van Buskirk, Cornejo, and Dalton (2008), college students played a military-based simulation called the Forward Observer PC-based Simulation (FOPCSIM) to train a call for fire task (see left panel of Fig. 7.3). The goal of this task was to destroy enemy targets before they get too close to the learner's position, paying attention to the priority of the targets. The task required three steps: the learner must locate the target, identify the target, and call the position and target type to the artillery or mortar unit. Students were assigned to one of four feedback conditions. In the percent accuracy feedback condition, students were told their overall score, such as "You correctly chose the high priority target 60 % of the time" (i.e., a type of outcome feedback). In the process feedback condition, students received information about how to perform the task, such as "Be sure to use the binoculars to scan the environment and refer to the prioritization rules" (i.e., informational feedback). In the normative feedback condition (i.e., a type of outcome feedback), students received information comparing their performance to that of others, but it was not specific, such as "You were in the 90th percentile. You performed better than 90 % of participants and 10 % performed better than you." Finally, in the control condition, students did not receive any feedback about how they performed the task. The authors found that the process feedback condition outperformed all of the other conditions on prioritization decisions (e.g., determining which targets were the highest priority for targeting actions) and that the normative and percent accuracy feedback conditions did not differ from the no-feedback control condition. They concluded that providing process feedback is an essential step for learning in simulation-based training environments because it focuses trainees on learning decision-making skills (e.g., steps to completing the task) versus performance outcomes (i.e., percent of correct responses).

7.3.1.3 Comparing Different Types of Process Feedback in a Military Procedural Learning Simulation

Billings (2012) and Serge, Priest, Durlach, and Johnson (2013) performed experiments to investigate the impact of different levels of feedback specificity on learning search and report procedures in a game-based simulation. In these experiments, college students learned about procedures for searching rooms for objects and personnel and the protocol for making reports to headquarters of what they found. Students then practiced these procedures in the game across three training scenarios and one performance scenario that was more complex with time pressure and low lighting conditions. The level of feedback specificity was manipulated between groups. In the general feedback condition, students received high-level feedback at the terminal learning objective level about their performance at the end of each training scenario (i.e., informational feedback), such as "Remember to apply the procedures for entering and exiting buildings." In the detailed feedback condition, students received more specific feedback at the enabling learning objective level about their performance at the end of each training scenario (i.e., topic specific feedback), such as, "Before entering or tagging a building, you should walk around the entire building to make sure it is not already tagged." In the outcome feedback condition, students received their performance score at the end of each training scenario (i.e., percent accuracy feedback). In the performance scenario, students in the detailed feedback condition outperformed those in the outcome feedback condition. In the Billings (2012) experiment, the general feedback condition performed similarly to the outcome feedback condition, and neither group showed any performance improvement across the four scenarios. To follow up on this finding, Serge et al. (2013) allowed students in the general feedback condition the option to review the training manual after receiving feedback but before starting a new training scenario (i.e., there were three opportunities to review the training manual) to remind themselves of the specific search and report procedures. The results showed that students who opted to review the manual performed at the same level as the detailed feedback condition, whereas those who did not review the manual performed as poorly as the outcome feedback condition. These results support the notion that students in the general feedback condition may have forgotten some of the procedures, and the high-level information provided in the feedback was not enough to help them repair their errors in subsequent scenarios. Therefore, both Billings (2012) and Serge et al. (2013) concluded that providing detailed feedback during training improves performance.

Taken together, the results of these experiments demonstrate that more detailed process feedback improves performance and learning outcomes for novice learners when compared to outcome feedback, consistent with the feedback principle in multimedia learning. Providing more specific feedback helps to reduce a learner's extraneous cognitive processing, by guiding them to the correct action or answer within the game, leading to higher learning outcomes and performance. Additionally, generative processing may be affected by detailed feedback in that this form of processing requires motivation to make sense of the material. Essential processing may not be affected by providing detailed feedback as feedback should not inherently change the difficulty of the material to be learned.

7.3.2 *Modality of Feedback*

The modality in which feedback is presented can have a large impact on the effectiveness of the feedback. The modality principle states that people learn more deeply when words are presented in spoken form rather than printed text in a primarily visual task (i.e., when the limited capacity of the visual channel is already occupied by visual information; Mayer, 2009, 2014a, 2014b). The theoretical rationale for the modality principle is that when learners are playing a visually based game and information is presented as printed text, their visual channel can become overloaded. By presenting that information in spoken form, the learner can offload some of the demands on the visual channel by utilizing the verbal channel to process the words. This allows learners to increase their cognitive processing capacity by utilizing both channels to process the incoming essential information; this is an example of managing essential processing.

The modality principle is well established in the multimedia learning literature (Ginns, 2005; Low & Sweller, 2014; Mayer & Moreno, 1998; Mayer & Pilegard, 2014; Mousavi, Low, & Sweller, 1995), and research evidence is beginning to show that it holds true for serious games. For example, Moreno, Mayer, Spires, and Lester (2001) performed a series of experiments examining the attributes of pedagogical agents that lead to deeper learning using the Design-A-Plant simulation game that was discussed earlier (Sect. 7.3.1.1). The authors examined how the modality of feedback and instructions delivered by an animated pedagogical agent (Experiment 4) or a human agent presented in the form of a video (Experiment 5) affects learning. One group received feedback and instructions from the agent in spoken form, while the other group received that information as onscreen text. Across both experiments, the authors found evidence for a modality effect, such that the spoken text group outperformed the onscreen text group on tests of retention and transfer and rated the material as more interesting. They concluded that students could devote more cognitive resources to meaningful learning by utilizing both the verbal and visual channels, thereby managing essential processing.

Using a more complex task, Fiorella, Vogel-Walcutt, and Schatz (2012) tested the modality principle in an interactive military simulation training a decision-making task. College students performed in two call-for-fire scenarios (a visual–spatial task), in which they had to decide the appropriate warning orders to give when firing on an enemy, taking into account the expense of ammunition chosen and level of damage to targets. The correct choice would be a warning order that costs less and is less destructive for vehicle targets or is more destructive and expensive for tank targets. Two groups of participants received real-time feedback on their performance either as onscreen text or spoken form, while a third group received no performance feedback, serving as a control. The authors found that the spoken feedback group performed better than the text feedback and control groups during the more complex training scenario and on an assessment scenario in which no feedback was provided. On retention tests of procedural knowledge, those in the spoken feedback group scored higher than the text feedback group, although there were no significant differences in scores for tests of declarative or integrative knowledge. The authors concluded that, consistent with CTML, when completing a visual–spatial task, real-time feedback should be spoken rather than presented as text to allow the learner to utilize both the verbal and visual channels to process the information. This frees up cognitive resources to engage in more productive cognitive processing to achieve meaningful learning. At the same time, providing real-time feedback as onscreen text overloads the learner's visual channel, leaving the learner with fewer cognitive resources to process the information, and he is unable to engage in the essential and generative processing necessary to achieve meaningful learning.

Another consideration to take into account when designing spoken feedback messages in serious games or simulations is the type of voice used: synthetic or natural speech. Synthetic speech is often reported in experiments as qualitatively different than natural speech (Paris, Thomas, Gilson, & Kincaid, 2000), leading individuals to encode the information differently. Likewise, Smither (1993) suggested that synthetic speech places more processing demands on a person than natural speech, which according to CTML would elicit extraneous cognitive processing on behalf of the learner and may lead to a learning decrement. In fact, Mayer and DaPra (2012) found that college students performed best on a transfer test when the animated pedagogical agent had a human voice versus a machine voice. They also found that other human-like features of the agent, such as eye gaze, facial expression, and body movements facilitated deeper learning, consistent with social agency theory.

In summary, the research on modality supports the idea that for a visual task, feedback is more effective when presented as spoken words (i.e., human speech). When playing a serious game with a highly visual component, presenting feedback as spoken words allows learners to utilize both their verbal and visual channels for processing the incoming information, thereby increasing their cognitive processing capabilities. Learners then have more cognitive resources available to devote to essential and generative processing, leading to better learning outcomes. However, this guidance does not suggest that feedback always be presented as spoken words, but merely to consider the nature of the game and the limits on a learner's cognitive system and to present feedback in the modality that is most appropriate.

7.3.3 Timing of Feedback

The timing of feedback presentation (i.e., immediate versus delayed) can also have an impact on learning and performance in serious games and simulations. However, guidance on when to give feedback is mixed due to the conflicting theories and research on the timing of feedback (Shute, 2008; Van Buskirk, 2011). For instance, those who argue for the use of immediate feedback suggest that it prevents errors from being encoded into memory (Bangert-Drowns et al., 1991). Temporal contiguity research suggests that when feedback is presented to a trainee immediately, the correct cue–strategy associations in the learning environment are strengthened and incorrect ones are weakened, thus resulting in better performance (Anderson, Corbett, Koedinger, & Pelletier, 1995; Corbett, Koedinger, & Anderson, 1997).

Alternative theories for feedback presentation suggest that when feedback is presented immediately after an error, the incorrect response interferes with learning the correct answer and learning the correct way to do the task (known as the perseveration-interference hypothesis). Conversely, errors made early on in learning may be forgotten if feedback is delayed, and the removal of "interference" increases the chances that learners will learn the correct information (Kulhavy & Anderson, 1972). Further, proponents of delayed feedback presentation also argue that immediate feedback serves as a crutch. In other words, learners come to rely on the feedback and when it is removed, their resulting performance suffers.

Kulik and Kulik (1988) performed a meta-analysis to compare the effectiveness of immediate and delayed feedback on verbal learning tasks in the laboratory and in the classroom. Their results showed that for applied studies, the use of immediate feedback yielded performance improvements over delayed feedback. However, results from laboratory studies showed the opposite effect, indicating an advantage of delayed feedback. In a recent study, Johnson, Priest, Glerum, and Serge (2013) investigated the timing of (detailed) feedback delivery in a game-based simulation for training search and report procedures that was discussed earlier (see Sect. 7.3.1.3). Participants received feedback according to three feedback timing schedule conditions: immediate (after an error), chunked (at a logical stopping point in the scenario), or delayed (at the end of a scenario). Although the authors found no statistically significant differences between the timing conditions, the trends in the data showed that the immediate feedback groups scored marginally higher than both the delayed and chunked groups. Further, they found that delayed feedback groups reported increases in cognitive load over time, while the chunked and immediate groups reported decreases in cognitive load. The authors concluded that providing immediate feedback may serve to reduce extraneous cognitive load for novices when learning procedures in a serious game. The results from this experiment are not overwhelming, yet they suggest some preliminary evidence for the use of immediate feedback in serious games. Certainly more research is needed to determine the optimal feedback timing schedules to enhance learning and performance in serious games and simulations.

In addition to the conflicting theoretical perspectives, the timing of feedback literature is riddled with inconsistencies due to several reasons. First, there is no agreement on the operational definitions of what is considered delayed versus immediate feedback. For example, most researchers would agree that feedback presented after a test item or a trial that lasts between 15 and 30 s is immediate feedback. Likewise, most researchers would agree that feedback presented days after performance would be considered delayed. However, researchers disagree on operational definitions in the middle ground. For instance, is feedback presented after an entire test (with multiple test items) or after the completion of a game or scenario that lasted 15 min (with several stimulus–response actions) considered immediate feedback or delayed feedback? Second, many of the studies in this domain contain confounds, such as providing a double stimulus exposure (Kulik & Kulik, 1988) and not controlling the amount of feedback given. For example, some researchers provide less feedback when given immediately, because they do not want to overload the participant during game performance. Last, the feedback content and other parameters, such as task characteristics, vary greatly in the feedback timing studies. Therefore, more systematic research is needed in this area to investigate the incremental effectiveness of the complex relationships between these feedback variables (e.g., content, timing, task parameters, modality, etc.). One experiment that combined feedback variables (Van Buskirk, 2011) investigated the relationship of feedback timing (immediate, delayed), content (outcome, process), and modality (visual, auditory) within the Forward Observer PC-based Simulation (FOPCSIM) described earlier as the Marine Corps call-for-fire task (a highly visual task; see left and right panel of Fig. 7.3). The results showed that participants who received immediate, auditory, process feedback outperformed all other groups on target prioritization performance. Studies taking a value-added approach to research such as this will help provide guidance for practitioners about what combinations of feedback parameters are effective for learning.

7.3.4 *Individual Differences*

Consideration for individual differences of the learner can lead to optimized feedback strategies when developing a serious game or simulation. Characteristics of individuals can affect the efficacy of feedback on learning outcomes and may interact with other individual differences or the type of feedback given (Narciss et al., 2014). Narciss et al. (2014) summarize factors that are typically addressed in feedback strategy research, including (a) prior knowledge, (b) motivation, (c) gender, and (d) meta-cognitive skills (e.g., self-assessment, help-seeking). Yet another way to characterize individual difference variables is to consider whether they are developed or stable learner characteristics (Vandewaeter, Desmet, & Clarebout, 2011). Developed learner characteristics, such as domain knowledge and task motivation, could change as a result of the learning process. More stable learner characteristics include gender, intelligence, spatial ability, and some have argued learning style (but see Pashler, McDaniel, Rohrer, & Bjork, 2008). Although there is a developing body of literature on individual differences in feedback, the research on how learner characteristics interact with feedback in serious games or training simulations remains sparse (Narciss et al., 2014; Timmers, Braber-van den Broek, & van den Berg, 2013).

Considering domain knowledge, a developed learner characteristic, evidence from multimedia learning research has shown that experts and novices differ in the amount of instructional support necessary to foster meaningful learning. The *expertise reversal effect* is the finding that instructional support that is helpful for novice learners may actually hinder learning of experts in that domain (Kalyuga, 2014; Kalyuga, Ayres, Chandler, & Sweller, 2003). There have been few studies examining this result in the feedback and gaming literatures in particular. One study by Smits, Boon, Sluijsmans, and van Gog (2008) investigated feedback content, timing, and learner characteristics (prior knowledge) on learning outcomes in a web-based environment teaching genetics. They varied feedback content by offering either global (i.e., outcome and informational feedback) or elaborated (i.e., outcome, informational, the worked-out solution, and topic specific feedback) information about students' answers. Feedback timing was also varied: presented either immediately following each task or delayed to every other task. The authors found no effect for feedback timing regardless of prior knowledge level. For learners with low prior knowledge, there was no effect of feedback content on learning outcomes; however, for learners with higher prior knowledge, those who received global feedback outperformed those who received elaborated feedback. This finding suggests that learners with higher prior knowledge of a task may need less detailed feedback.

Considering spatial ability, a generally more stable learner characteristic, Moreno and Mayer (1999) found that the effectiveness of process feedback was influenced by the learner's spatial ability. In their experiment, elementary school children learned addition and subtraction playing an interactive multimedia program. For students who received process feedback, those with higher spatial ability had better learning gains from pre- to posttest than those with lower spatial ability. The authors reasoned that this difference in the effectiveness of process feedback is due to learners with lower spatial ability being cognitively overloaded during learning and not having the cognitive resources necessary to process the information in the feedback. On the other hand, the students with high spatial ability did have the cognitive resources available to process the feedback, and as a result they demonstrated higher learning gains. This finding supports the notion that spatial ability may be an important learner characteristic to consider when developing interactive learning environments for teaching complex skills, but certainly more research is needed to fully explore the relationships between spatial ability and type of feedback in various contexts.

Considering gender, another stable learner characteristic, Landsberg et al. (2010) investigated the relative efficacy of process (i.e., topic specific), velocity, and no feedback in FOPCSIM, the call for fire simulation (see left panel of Fig. 7.3). In this study, the results showed that gender significantly interacted with feedback condition on target prioritization performance, meaning that the most useful feedback was different for males and females. Specifically, they found that males performed better (i.e., correctly selected the highest priority target) when they received velocity feedback compared to process feedback, although males in the no feedback control group were not significantly different from either process or velocity feedback groups. Conversely, females performed significantly better when they received process feedback as compared to no feedback. These results suggest that process feedback helped females more accurately prioritize, while velocity feedback helped males more accurately prioritize. Furthermore, they found that females who received process feedback significantly outperformed males who received process feedback on target prioritization. This research provides evidence for the notion that a stable learner characteristic, such as gender, may influence the effectiveness of process feedback in serious games and simulations.

Finally, Narciss et al. (2014) sought to determine the effect and interaction of both stable and developed learner characteristics (e.g., gender, intrinsic motivation) on the efficacy of feedback on problem-solving behavior of sixth and seventh grade boys and girls in a computer adaptive math-learning program. The results showed mixed effects and interactions of learner characteristics with feedback content on learning and behavior outcomes. Overall, they found that there were differences between boys and girls in how much feedback impacted learning outcomes, such that boys benefited less from feedback on practice tasks, had lower knowledge gains from pre- to posttest, and their perceived competence declined. In contrast, girls did have significant gains on the knowledge posttest.

Taken together, these studies provide evidence that serious games designers and researchers should consider relevant individual differences for specific educational and training domains and how those learner characteristics may interact with different feedback strategies. More specifically, stable learner characteristics, such as gender and spatial ability, may have an impact on the efficacy of process feedback in serious games. Additionally, there is some evidence that developed learner characteristics may also impact the effectiveness of process feedback, such as level of domain expertise, but more systematic research is needed to understand which variables may play a role in which domains.

7.4 Discussion

7.4.1 Summary of Findings

In this chapter, we assessed the feedback literature in serious games from a value-added perspective, identifying key findings to arrive at the most applicable feedback characteristics benefiting learning outcomes. The majority of the studies address the effects of feedback content, modality, timing, and/or adaptation to individual differences as possible strategies for better learning through serious games. For each of these feedback features, we summarize the findings below:

1. *Content*

   The content of feedback refers to the type of information presented to learners. Feedback content may be combined in numerous ways depending on the learning objectives and learner characteristics. We grouped these feedback types into outcome-based feedback (informing learners about their performance outcome or learning progress) or process-based feedback (directing learners on the process to reach the correct answer or progress). Overall, research on feedback content seems to suggest that process (explanatory) feedback is more beneficial than outcome (corrective) feedback, because providing a deeper explanation that guides learners to the correct answer and reasoning behind it rather than merely informing learners about the correctness of their performance reduces extraneous processing and may enhance essential processing, leading to higher learning outcomes. However, there is some evidence that individual differences may impact efficacy of process feedback (see below).

2. *Modality*
   Feedback modality is the way in which the feedback information is presented. For example, feedback could be visual or auditory/spoken, delivered by a pedagogical agent or video of a person, and many other forms of presentation. Although conclusions at this point are tenuous as these studies vary widely in terms of educational domain and type of game or simulation, determining the best modality for feedback presentation appears to depend largely on the design of the instructional material. Good design should take into consideration the amount of cognitive load placed on the learner such that the modality of feedback does not compete with the instructional material for limited mental resources. That is, for a highly visual–spatial game, presenting feedback as spoken text allows learners to utilize both the visual and verbal channels to process the information.
3. *Timing*
   The literature on timing of feedback in serious games is sparse, but a trend may be developing to indicate an advantage for immediate feedback for acquisition of learning material, although how feedback timing affects long-term retention is still unclear. Further, feedback timing research indicates that this variable may have complex relationships with other feedback features (i.e., content and modality) and should not be investigated in isolation.
4. *Adaptation to Individual Differences*
   In general, the interaction between feedback strategies and characteristics of the individual learner is not well known. However, several research studies suggest that stable learner characteristics (e.g., gender, spatial ability) may impact the effectiveness of different feedback types. Although results on the relationship between developed learner characteristics and feedback variables are scant, initial findings suggest that level of prior knowledge can impact the effectiveness of process feedback.

7.4.2 Future Research Directions and Conclusions

This review highlights the lack of systematic research on feedback in serious games necessary to define the parameters of feedback strategies that optimize learning outcomes. The studies described throughout this chapter demonstrate how the literature on feedback in simulations and games varies greatly on terminology used, characteristics of feedback given, educational or training domain, types of learning outcomes, and the game itself. Further, the differing feedback features are often not described in enough detail to support replication and a comparison of results between studies (Hatala, Cook, Zendejas, Hamstra, & Brydges, 2014; Ke, 2009; Merchant et al., 2014). Therefore, authors should be mindful to provide examples and explicit details of the feedback and task they use in their research so that future researchers can make comparisons and draw conclusions from the large bodies of research in serious games and simulations. Additionally, in order to develop evidenced-based principles and guidelines for practitioners, we call for a value-added approach to future research in this domain to understand the relationships between feedback variables (content, modality, and timing) and learner characteristics (developed and stable). Lastly, many of the research findings presented were based on performance improvements during acquisition (with the exception of Moreno, 2004; Mayer & Moreno, 2005). However, future research should focus on long-term retention and transfer. This is not a new concept (see Schmidt & Bjork, 1992), but one that has ramifications for research in this area. The goal of creating educational and training materials is to produce lasting effects on knowledge, skills, or behavior, and the focus of current feedback literature on acquisition as opposed to retention limits the conclusions we can draw from studies of feedback parameters. In fact, a study by Astwood and Smith-Jentsch (2010) found that while immediate feedback groups had higher performance improvements during acquisition, the complete opposite results were found showing an advantage for delayed feedback during a retention phase.

Accordingly, more research is needed to determine the effects of feedback on learners' motivation when playing serious games. Although many researchers extol the virtues of games for promoting motivation, there is little empirical evidence to validate these claims. One recent meta-analysis even concluded that playing serious games (in general) may not be more motivating to students than traditional teaching methods (Wouters et al., 2013). We recommend that future researchers take a value-added approach to studying the effects of feedback (and other instructional strategies) on motivation and learning in serious games. Such research is necessary to develop an understanding of the mechanism by which motivation affects cognitive processing, which instructional strategies are most effective at fostering motivation, and whether particular individual difference variables enhance (or diminish) motivation.

Given that a learner's cognitive system has limited capacity, designers should be mindful of how feedback variables impose processing demands. From a review of the literature, we were able to identify features of feedback that could reduce cognitive processing demands to achieve improved learning and performance in serious games and simulations: feedback content, modality, timing, and adaptation to individual differences. However, there is still a wealth of systematic research that is needed in order to provide evidenced-based design principles for serious games and simulation designers.

References

Anderson, J. R., Corbett, A. T., Koedinger, K., & Pelletier, R. (1995). Cognitive tutors: Lessons learned. *The Journal of the Learning Sciences, 4*, 167–207.

Astwood, R., & Smith-Jentsch, K. (2010, April). *Feedback timing in team training: Moderating effects of goal orientation*. Poster presented at the 25th Annual Meeting of the Society for Industrial and Organizational Psychology, Atlanta, GA.

Astwood, R. S., Van Buskirk, W. L., Cornejo, J. M., & Dalton, J. (2008). The impact of different feedback types on decision-making in simulation based training environments. *Proceedings of*

Human Factors and Ergonomics Society Annual Meeting (Vol. 52, No. 26, pp. 2062–2066). Santa Monica, CA: Sage.

Azevedo, R., & Bernard, R. M. (1995). A meta-analysis of the effects of feedback in computer-based instruction. *Journal of Educational Computing Research, 13*(2), 11–127.

Bangert-Drowns, R. L., Kulik, C. L. C., Kulik, J. A., & Morgan, M. T. (1991). The instructional effect of feedback in test-like events. *Review of Educational Research, 61*(2), 213–238.

Billings, D. R. (2012). Efficacy of adaptive feedback strategies in simulation-based training. *Military Psychology, 24*, 114–133.

Cameron, B., & Dwyer, F. (2005). The effect of online gaming, cognition, and feedback type in facilitating delayed achievement of different learning objectives. *Journal of Interactive Learning Research, 16*(3), 243–258.

Corbett, A. T., Koedinger, K. R., & Anderson, J. R. (1997). Intelligent tutoring systems. In M. G. Helander, T. K. Landauer, & P. V. Prabhu (Eds.), *Handbook of human–computer interaction* (pp. 849–874). Amsterdam: Elsevier.

Fiorella, L., Vogel-Walcutt, J. J., & Schatz, S. (2012). Applying the modality principle to real-time feedback and the acquisition of higher-order cognitive skills. *Educational Technology Research and Development, 60*, 223–238.

Garris, R., Ahlers, R., & Driskell, J. E. (2002). Games, motivation, and learning: A research and practice model. *Simulation & Gaming, 33*, 441–466.

Gee, J. P. (2007). *What video games have to teach us about learning and literacy* (2nd ed.). New York: Palgrave MacMillan.

Ginns, P. (2005). Meta-analysis of the modality effect. *Learning and Instruction, 15*, 313–331.

Hatala, R., Cook, D. A., Zendejas, B., Hamstra, S. J., & Brydges, R. (2014). Feedback for simulation-based procedural skills training: A meta-analysis and critical narrative synthesis. *Advances in Health Science Education, 19*, 21–272.

Hattie, J., & Timperley, H. (2007). The power of feedback. *Review of Educational Research, 77*, 81–112.

Hays, R. T. (2005). *The effectiveness of instructional games: A literature review and discussion* (Technical Report 2005-004). Orlando, FL: Naval Air Warfare Center Training Systems Division.

Johnson, C. I., & Priest, H. A. (2014). The feedback principle in multimedia learning. In R. E. Mayer (Ed.), *The Cambridge handbook of multimedia learning* (2nd ed., pp. 449–463). New York: Cambridge University Press.

Johnson, C. I., Priest, H. A., Glerum, D. R., & Serge, S. R. (2013). Timing of feedback delivery in game-based training. *Proceedings of the Interservice/Industry Training, Simulation & Education Conference*, Orlando, FL, 2013. Arlington, VA: National Training Systems Association.

Kalyuga, S. (2014). The expertise reversal principle in multimedia learning. In R. E. Mayer (Ed.), *The Cambridge handbook of multimedia learning* (2nd ed., pp. 576–597). New York: Cambridge University Press.

Kalyuga, S., Ayres, P., Chandler, P., & Sweller, J. (2003). The expertise reversal effect. *Educational Psychologist, 38*, 23–31.

Ke, F. (2009). A qualitative meta-analysis of computer games as learning tools. In R. E. Ferdig (Ed.), *Handbook of research on effective electronic gaming in education* (Vol. 1, pp. 1–32). Hershey, PA: Information Science Reference.

Kirschner, P. A., Sweller, J., & Clark, R. E. (2006). Why minimal guidance during instruction does not work: An analysis of the failure of the constructivist, discovery, problem-based, experiential, and inquiry-based teaching. *Educational Technologist, 41*, 75–86.

Kluger, A. N., & DeNisi, A. (1996). The effects of feedback interventions of performance: A historical review, a meta-analysis, and a preliminary feedback intervention theory. *Psychological Bulletin, 119*(2), 254–284.

Kulhavy, R. W., & Anderson, R. C. (1972). Delay-retention effect with multiple-choice tests. *Journal of Educational Psychology, 63*, 505–512.

Kulik, J. A., & Kulik, C. L. C. (1988). Timing of feedback and verbal learning. *Review of Educational Research, 58*(1), 79–97.

Landsberg, C., Van Buskirk, W. L., & Astwood, R. S. (2010). Does feedback type matter? Investing the effectiveness of feedback content on performance outcomes. *Proceedings of the 54th Annual Meeting of the Human Factors and Ergonomics Society*, San Francisco, CA.

Low, R., & Sweller, J. (2014). The modality principle in multimedia learning. In R. E. Mayer (Ed.), *The Cambridge handbook of multimedia learning* (2nd ed., pp. 227–246). New York: Cambridge University Press.

Mason, B., & Bruning, R. (2003). *Providing feedback in computer-based instruction: What the research tells us.* Retrieved from http://dwb.unl.edu/Edit/MB/MasonBruning.html

Mayer, R. E. (2003). Should there be a three-strikes rule against discovery learning? The case for guided methods of instruction. *American Psychologist, 59*, 14–19.

Mayer, R. E. (2009). *Multimedia learning* (2nd ed.). New York: Cambridge University Press.

Mayer, R. E. (2014a). *Computer games for learning: An evidence-based approach.* Cambridge, MA: MIT Press.

Mayer, R. E. (2014b). *The Cambridge handbook of multimedia learning* (2nd ed.). New York: Cambridge University Press.

Mayer, R. E., & DaPra, C. S. (2012). An embodiment effect in computer-based learning with animated pedagogical agents. *Journal of Experimental Psychology: Applied, 18*, 239–252.

Mayer, R. E., & Johnson, C. I. (2010). Adding instructional features that promote learning in a game-like environment. *Journal of Educational Computing Research, 42*, 241–265.

Mayer, R. E., & Moreno, R. (1998). A split-attention effect in multimedia learning: Evidence for dual processing systems in working memory. *Journal of Educational Psychology, 90*, 312–320.

Mayer, R. E., & Pilegard, C. (2014). Principles for managing essential processing in multimedia learning: Segmenting, pre-training, and modality principles. In R. E. Mayer (Ed.), *The Cambridge handbook of multimedia learning* (2nd ed., pp. 316–344). New York: Cambridge University Press.

Merchant, Z., Goetz, E. T., CiFuentes, L., Keeney-Kennicutt, W., & Davis, T. J. (2014). Effectiveness of virtual reality-based instruction on students' learning outcomes in K-12 and higher education: A meta-analysis. *Computers & Education, 70*, 29–40.

Moreno, R. (2004). Decreasing cognitive load for novice students: Effects of explanatory versus corrective feedback in discovery-based multimedia. *Instructional Science, 32*, 99–113.

Moreno, R., & Mayer, R. E. (1999). Multimedia-supported metaphors for meaning making in mathematics. *Cognition and Instruction, 17*(3), 215–248.

Moreno, R., & Mayer, R. E. (2005). Role of guidance, reflection, and interactivity in an agent-based multimedia game. *Journal of Educational Psychology, 97*, 117–128.

Moreno, R., & Mayer, R. E. (2007). Interactive multi-modal environments: Special issue on interactive multimodal environments- Contemporary issues and trends. *Educational Psychology Review, 19*, 309–326.

Moreno, R., Mayer, R. E., Spires, H. A., & Lester, J. C. (2001). The case for social agency in computer-based teaching: Do students learn more deeply when they interact with animated pedagogical agents? *Cognition and Instruction, 19*, 177–213.

Mousavi, S., Low, R., & Sweller, J. (1995). Reducing cognitive load by mixing auditory and visual presentation modes. *Journal of Educational Psychology, 87*, 319–334.

Narciss, S., & Huth, K. (2004). How to design informative tutoring feedback for multi-media learning. In H. M. Niegemann, R. Brünken, & D. Leutner (Eds.), *Instructional design for multimedia learning* (pp. 181–195). Münster: Waxmann.

Narciss, S., Sosnovsky, S., Schnaubert, L., Andrès, E., Eichelmann, A., Goguadze, G., et al. (2014). Exploring feedback and student characteristics relevant for personalizing feedback strategies. *Computers and Education, 71*, 56–76.

O'Neil, H. F., & Perez, R. S. (Eds.). (2008). *Computer games and team and individual learning.* Amsterdam: Elsevier.

Paris, C. R., Thomas, M. H., Gilson, R. D., & Kincaid, J. P. (2000). Linguistic cues and memory for synthetic and natural speech. *Human Factors: The Journal of the Human Factors and Ergonomics Society, 42*(3), 421–431.

Pashler, H., McDaniel, M., Rohrer, D., & Bjork, R. (2008). Learning styles: Concepts and evidence. *Psychological Science in the Public Interest, 9*, 105–119.

Prensky, M. (2001). *Digital game-based learning*. New York: McGraw-Hill.

Schmidt, R. A., & Bjork, R. A. (1992). New conceptualizations of practice: Common principles in three paradigms suggest new concepts for training. *Psychological Science*, 207-217.

Serge, S. R., Priest, H. A., Durlach, P. J., & Johnson, C. I. (2013). The effects of static and adaptive performance feedback in game-based training. *Computers in Human Behavior, 29*, 1150–1158.

Shute, V. J. (2008). Focus on formative feedback. *Review of Educational Research, 78*, 153–189.

Sitzmann, T. (2011). A meta-analytic examination of the instructional effectiveness of games for computer-based simulation games. *Personnel Psychology, 64*, 489–528.

Smither, J. A. A. (1993). Short term memory demands in processing synthetic speech by old and young adults. *Behaviour & Information Technology, 12*, 330–335.

Smits, M. H. S. B., Boon, J., Sluijsmans, D. M. A., & van Gog, T. (2008). Content and timing of feedback in web-based learning environment: Effects on learning as a function of prior knowledge. *Interactive Learning Environments, 16*(2), 183–193.

Timmers, C. F., Braber-van den Broek, J., & van den Berg, S. (2013). Motivational beliefs, student effort, and feedback behavior in computer-based formative assessment. *Computers & Education, 60*, 25–31.

Van Buskirk, W. L. (2011). *Investigating the optimal presentation of feedback in simulation-based training: An application of the cognitive theory of multimedia learning*. Unpublished doctoral dissertation, University of Central Florida, Orlando.

Vandewaeter, M., Desmet, P., & Clarebout, G. (2011). The contribution of learner characteristics in the development of computer-based adaptive learning environments. *Computers in Human Behavior, 27*, 118–130.

Vogel, J. J., Vogel, D. S., Cannon-Bowers, J., Bowers, C. A., Muse, K., & Wright, M. (2006). Computer gaming and interactive simulations for learning: A meta-analysis. *Journal of Educational Computing Research, 34*, 229–243.

Wouters, P., van Nimwegen, C., van Oostendorp, H., & van der Speck, E. D. (2013). A meta-analysis of the cognitive and motivational effects of serious games. *Journal of Educational Psychology, 105*, 249–265.

Wouters, P., & van Oostendorp, H. (2013). A meta-analytic review of the role of instructional support in game-based learning. *Computers & Education, 60*, 412–425.

Chapter 8
Self-Explanations in Game-Based Learning: From Tacit to Transferable Knowledge

Judith ter Vrugte and Ton de Jong

Abstract Game-based learning is often considered to be an effective instructional approach, but the effects of game-based learning are varied and far from optimal. Aside from many features and characteristics that might affect the results of game-based learning, we conjecture that games generally thrive on experiential learning and that experiential learning does increase knowledge, but that this knowledge is often implicit. We note that though implicit knowledge is certainly valuable, that in general explicit knowledge is considered more desirable in education, because it is more accessible and promotes transfer. It is suggested that explicit knowledge does not always automatically follow from the development of implicit knowledge, but that this process can be supported through self-explanations. Because self-explanations rarely occur automatically in game-based learning environments, we propose that self-explanations in game-based learning environments can be elicited by specific instructional approaches. Three possible approaches for eliciting self-explanations are discussed: question prompts, collaboration, and partial worked examples.

Keywords Self-explanation • Question prompt • Collaboration • Partial worked example

8.1 Introduction

As society develops, education also develops. The quill pen was replaced by the fountain pen and ballpoint, and slates have been replaced by spiral notebooks. And today, smartphones, tablets, and laptops with touchscreen, keyboard, and mouse are common sights in everyday education. This introduction of technology into the classroom opened up opportunities for implementation of alternative teaching strategies in education. What, in turn, stimulated the introduction of computer games as an educational tool, and increased the relevance, significance, and

J. ter Vrugte (✉) • T. de Jong
Department of Instructional Technology, University of Twente,
PO Box 217, 7500 AE Enschede, The Netherlands
e-mail: j.tervrugte@utwente.nl

P. Wouters, H. van Oostendorp (eds.), *Instructional Techniques to Facilitate Learning and Motivation of Serious Games*, Advances in Game-Based Learning, DOI 10.1007/978-3-319-39298-1_8

influence of computer game-based learning research. Although computer games potentially provide a medium for high quality learning, the learning does not seem to come automatically (Girard, Ecalle, & Magnan, 2013; ter Vrugte & de Jong, 2012). As with all complex multimedia learning environments, many students experience difficulties with appropriately processing the static and dynamic media employed in such environments (for an overview, see Mayer, 2005). In the current chapter, we discuss different ways to elicit self-explanation to help students effectively process the educational content in educational games.

8.2 Theoretical Position or Motivation

8.2.1 When Games Enter the Classroom

Motivation is a highly valued characteristic of educational games; for this reason, most game developers design games in which students feel as though they are playing instead of learning. This can influence the students' learning mode (i.e., their level of intentionality): instead of a state of deliberative learning (i.e., intentional and conscious learning (Eraut, 2000)) they adopt a state in which learning is reactive (i.e., near-spontaneous and unplanned (Eraut, 2000)), or even implicit (i.e., 'in the absence of explicit knowledge about what was learned' (Reber, 1993, p. 5)). The 'learning mode' can affect specific features of the knowledge that is developed. In general, it seems that when the learning mode becomes less intentional, the development of explicit knowledge (knowledge that can be articulated) becomes less likely (Eraut, 2000; Reber, 1993).

In addition, most games capitalize on experiential learning; they promise to engage and motivate students through direct experiences with the game world (Kiili, 2005). While students would typically learn in a top-down approach—receive explicit knowledge through instruction and proceduralize this knowledge through practice—experiential learning generally follows a bottom-up approach: students acquire knowledge through experience and practice (Eraut, 2000; Sun, Merrill, & Peterson, 2001). As a consequence of this experiential approach to learning, the learning becomes more intuitive and implicit.

Research on implicit learning has demonstrated that implicit knowledge is not always accompanied by explicit knowledge and vice versa (Berry & Broadbent, 1984). In implicit and reactive learning modes, students are more likely to obtain implicit knowledge; the knowledge gathered is therefore often tacit, rather than explicit (Eraut, 2000; Reber, 1993). In a study specifically about knowledge gain in game-based learning, Leemkuil and de Jong (2012) found no correlation between knowledge gain and game performance. Students gained implicit knowledge (improved performance during the game), but this gain did not translate into a gain in explicit knowledge (i.e., improved performance on knowledge tasks/transfer tasks).

Though implicit knowledge is valuable and measurable, explicit knowledge is generally our goal, because it is this explicit knowledge that increases recall and accessibility and promotes transfer (Wouters, Paas, & van Merriënboer, 2008). This, in turn, enables students to deploy their knowledge in more than one context, and fosters the ability to communicate it to others (Sun et al., 2001). In addition, school tests are commonly designed to evaluate explicit knowledge, and only occasionally directly measure implicit knowledge. Therefore, when a game relies on implicit learning and as a result only improves implicit knowledge, students and teachers might fail to see the value of playing the educational game. And in some cases, because learning content is so intertwined with game-content, students and teachers even fail to see the connection between the game activities and the curricular content (Barzilai & Blau, 2014).

8.2.2 Self-Explanations to Foster Game-Based Learning

From the discussion thus far, we can identify several problems that arise with the introduction of game-based learning. The problems seem to derive from the learning mode and learning process that can be associated with game-based learning, which generally stimulate the development of procedural knowledge and skills rather than explicit knowledge (Leemkuil & de Jong, 2012). Finding a way to stimulate the development of explicit knowledge in game-based learning would make educational games more useful, because the connection between the game activities and the educational curriculum and learning objectives of the school would be more evident. And, most importantly, explicit knowledge fosters transfer, enabling students to reproduce the knowledge and put it into practice.

In order to construct explicit knowledge, students must be aware of what they are doing and how they are doing it. This awareness is facilitated by self-explanation. Self-explanation is "a constructive activity that engages students in active learning, and ensures that students attend to the material in a meaningful way" (Roy & Chi, 2005, p. 273). It is a process of conscious reflection on, and analysis of, the output generated by implicit knowledge (Boud, Keogh, & Walker, 1985; Jordi, 2010). Self-explanation has been found to be an essential element in learning (Barab, Thomas, Dodge, Carteaux, & Tuzun, 2005; Ke, 2008), and more specifically, in experiential learning (Jordi, 2010). It has been demonstrated that the more students self-explain, the more they learn. In studies that focus on learning from worked examples, this is referred to as 'the self-explanation effect' (Chi, Bassok, Lewis, Reimann, & Glaser, 1989; Chi, De Leeuw, Chiu, & Lavancher, 1994). In a review, Roy and Chi (2005) report that self-explanation results in learning gains as much as 44 % higher than gains for control conditions without self-explanations. However, when playing a game students can be reluctant to take the time to think about their actions and reflect on the outcomes, due to the phenomenon of game flow (Ke, 2008). Students keep experimenting until their scores improve, but this trial-and-error behavior rarely enhances explicit knowledge (Kiili, 2005). When we design educational games that intend to capitalize on experiential, implicit learning (bottom-up learning), we should find a way to foster self-explanations if we want to optimize the effectiveness of the games.

This is easier said than done. As mentioned before, games capitalize on their motivational appeal. Any alterations can affect the experience of game flow and diminish motivational effects. Therefore, any modifications that are designed to turn playing into learning need to be implemented with great care. As Killingsworth, Clark, and Adams (2015, p. 62) justly point out, "implementing self-explanation in educational games requires careful consideration of the specific affordances and constraints of digital games as a medium, and careful evaluation of the relationship between individual abilities, gameplay, and learning outcomes."

In the following sections, we discuss possible means of initiating self-explanation to stimulate the generation of explicit knowledge and thereby optimize learning from educational games. We present three promising options that could support the generation of explicit knowledge by stimulating self-explanation: self-explanation prompts, collaboration, and worked-out examples. We also discuss the implications for game-based learning research and design.

8.3 Self-Explanation Prompts

8.3.1 Categorization of Self-Explanation Prompts

Research shows that self-explanation is effective in many learning domains (see Wylie and Chi 2014, for an overview). And though most students are likely to engage in some form of spontaneous self-explanation, the quality and quantity of these explanations varies. For this reason, researchers have investigated ways to prompt self-explanations (to increase their quantity) and also to support them (to increase their quality). Wylie and Chi (2014) introduced the 'continuum of different forms of self-explanation,' by which they categorize different forms of prompted self-explanations. The categorization is based on the level of structure that the prompts and scaffolds provide, and is therefore related to the level of cognitive processing. In general, we expect that when support increases, processing demands decrease.

Prompts can be open, meaning that they indicate that the student should self-explain, but give no information about the content/focus of the self-explanation. Alternatively, prompts can be directive (focused), meaning that the prompts contain information about the content/focus of the self-explanation. Directive prompts could be question prompts or fill-in-the-blank prompts. For example, in directive question prompts (e.g., 'Why did the shadow lengthen?') students have to phrase (no support) or select (support) an answer. In directive fill-in-the-blank prompts (i.e., When the height of the light source _______ the shadow lengthens.) students have to complete a statement/explanation by typing (no support) or selecting (support) the correct word(s). The support that students receive when expressing their reaction can also vary. We can discriminate between 'source-based' and

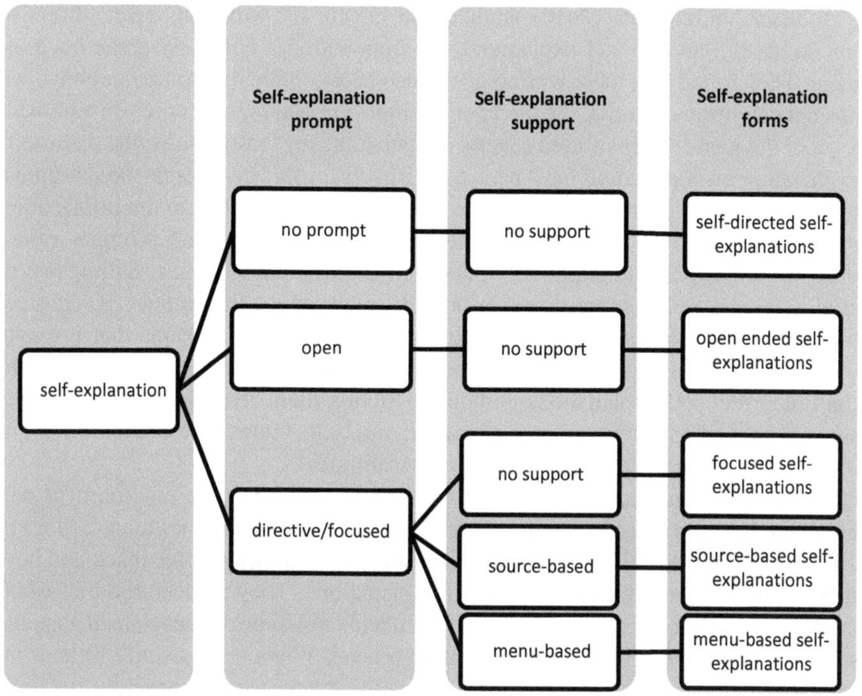

Fig. 8.1 Categorization of self-explanations based on 'the continuum of forms of self-explanation' by Wylie and Chi (2014)

'menu-based' types of support. Source-based means that students can select their response from a library in which the options are limited to a specific domain, but are not specifically matched to the prompt. Students receive some support and therefore the processing demands decrease. Menu-based means that students can select their answer from a list of options, and the options are matched to the prompt (like a multiple choice answer). With menu-based support students receive more structure and focus, and therefore processing demands are likely to be even lower. We provide an overview of the types of self-explanations, organized from highest processing demands to lowest, in Fig. 8.1.

The effectiveness of self-explanation prompts can be affected by many factors: characteristics of the prompts, characteristics of the students, and characteristics of the environments in which the prompts are placed. As shown in Fig. 8.1, at the level of the prompts we can differentiate between directive or open prompts. Both directive and open self-explanation prompts have advantages and disadvantages. Directive prompts may restrict students and can thus limit students' opportunities to learn (Chi, 2000), but open prompts may be too demanding and result in extraneous processing (O'Neil et al., 2014). In addition, without support, some students might not be capable of responding appropriately to self-explanation prompts (Berthold, Eysink, & Renkl, 2009).

Johnson and Mayer (2010) studied the effects of providing open, directive (menu-based), and no self-explanation prompts within a computer game environment. They found that there was no difference between the no prompt and the open prompt conditions, but that the directive prompts yielded higher scores on a transfer level of the game. They offered several explanations for these results: the open self-explanation prompts could have been too difficult for the students or too disruptive to the game flow, or the observed effects might have resulted from the information that was provided by the multiple choice answers in the directive prompts, rather than the self-explanations per se. In addition to these explanations, Killingsworth et al. (2015) explained that directive prompts may reduce the amount of incorrect thinking, and therefore also reduce extraneous processing. It seems that prompts will be most effective when they provide just enough structure to guide students and limit incorrect self-explanations without restricting them. To achieve this, prompts should be matched to the students' characteristics (e.g., knowledge about the domain in question, and cognitive and metacognitive abilities).

Besides student characteristics, characteristics of the game environment can affect the benefits of adding self-explanation prompts. In two experiments, Moreno and Mayer (2005) sought to identify the role of interactivity and feedback and how these influence the effectiveness of self-explanations. They conjectured that when tasks are highly interactive, meaning that students must make conscious decisions, students' cognitive activity is already at high level. When tasks require little or no interactivity, where meaning is constructed without engaging in conscious decision making, students' cognitive activity would be at a lower level. Thus, because interactivity in itself already fosters processes comparable to self-explanations, addition of self-explanation prompts should be more helpful in noninteractive environments than in interactive environments. Results of these experiments indicated that self-explanation promotes retention and transfer in noninteractive environments, but not in interactive ones.

8.3.2 Accuracy of Self-Explanations

In a series of experiments, Hsu and Tsai looked into the effects of directive (menu-based) self-explanations (Hsu & Tsai, 2011, 2013; Hsu, Tsai, & Wang, 2012). They implemented self-explanation prompts in a game-based environment designed to teach elementary school children about light and shadow concepts. In the game (Hsu et al., 2012; Hsu & Tsai, 2011, 2013), students had to adjust the angle of a flashlight to keep the shadow of their avatar within a specified field. The prompts were multiple choice questions that appeared whenever the player made a mistake, and prompted the student to specify the possible cause for their failure. Hsu and Tsai (2011, 2013) found that in terms of conceptual understanding (posttest with ten multiple choice questions about light-shadow concepts), students who played the game with the directive self-explanation prompts did not outperform those who played the game without the prompts. In a replication study, Hsu et al. (2012) added a retention test. Results of this study indicated no difference between the self-explanation condition and the control condition on posttest and retention test. However, they did find that the students in the self-explanation condition who engaged effectively with the self-explanation prompts (over 50 % of their provided explanations were correct) showed a significantly higher score on the retention test than students whose engagement with the self-explanation prompts was less successful (less than 50 % of the provided self-explanations were correct) and students in the control condition.

In a follow-up study, Hsu, Tsai, and Wang (in press) replicated their earlier study (i.e., Hsu et al., 2012) and added a condition in which students played a multiplayer (two player) version of the game. In this condition, students would receive a directive (menu-based) self-explanation prompt not only when they made a mistake, but also when their partner made a mistake. Both students had to respond to the prompt correctly before they could continue playing the game. When one of the two students responded incorrectly, the other student had to provide advice by indicating the correct response to the multiple choice question (menu-based self-explanation). Thus, students in the multiplayer condition were more likely to receive a self-explanation prompt and would receive a suggestion from the other student when their response to the prompt was incorrect. Results indicated that students in the individual, self-explanation condition outperformed the students in the multiplayer self-explanation condition and control conditions on the posttest, but there were no differences between the conditions on retention test scores. Exploration of the relation between the accuracy of the self-explanations and test performance showed no clear relation between accuracy and posttest performance. However, in line with their earlier findings (Hsu et al., 2012) there was a clear relation between accuracy of the self-explanations and retention test performance. On the retention test, regardless of condition highly accurate students outperformed students with low accuracy and students in the control group. We emphasize that in both studies (Hsu et al. 2012, in press) accuracy only affected the retention test scores, not the posttest scores. Other studies that explored the relation between the accuracy of responses in the self-explanation condition, and posttest performance found no differences between students with high accuracy and students with low accuracy (Adams & Clark, 2014; Hsu & Tsai, 2013). All of these studies employed menu-based self-explanations, and the posttest was completed directly after the last game session (i.e., Adams & Clark, 2014; Hsu et al., 2012; Hsu & Tsai, 2013).

Though the previous studies show that the accuracy of students' reactions to self-explanation prompts might affect the retention of knowledge, they provide neither insight into how students' abilities and prior knowledge influence this effect, nor do they indicate whether specific characteristic of the prompts might elicit more accurate or effective self-explanations. O'Neil et al. (2014) investigated an educational game with directive self-explanation prompts and noticed that not all prompts elicited equally effective self-explanations. The effect of the self-explanations depended on the content and focus of the prompt. They differentiated between three types of directive (menu-based) self-explanation prompts: recall (essential processing), focused (essential and generative processing), and abstract (generative processing).

The recall prompts were simple questions designed to focus students' attention on the relevant information. The focused prompts were also relatively simple questions designed to help students see the relation between the game elements and educational content, and the abstract prompts were questions designed to encourage students to consider the reasons for specific actions and outcomes. Results of their study show that students benefitted most from the focused self-explanation prompts. O'Neil et al. (2014) concluded that self-explanation prompts were most effective when they helped students to see the relation between the game terminology and (in their case) mathematics terminology. Based on their study, we can conclude that it is not only the form (see Fig. 8.1), but also the focus that can affect the level of cognitive processing elicited by the self-explanation prompts.

8.3.3 Self-Explanation vs Feedback

The studies just discussed show that self-explanation prompts are not always effective and indicate that their effectiveness could depend on the quality of the self-explanations elicited, in that students who generate more accurate explanations tend to be more likely to generate sustainable knowledge. So why would we favor self-explanation, with the risk of incorrect self-explanations, over providing students with correct explanations (explanatory feedback)? One of the most salient differences between the two options is the level of student engagement. While self-explanation requires students to engage actively, explanatory feedback does not (though we must note that students can use self-explanations to interpret the feedback). In general, it is assumed that when students' engagement increases, their learning increases (Chi & Wylie, 2014; Renkl, 2002; Roy & Chi, 2005). Some researchers have explored the differences between providing students with accurate, explanatory feedback and eliciting self-explanations.

Moreno and Mayer (2005) compared feedback and *open-ended self-explanation* and demonstrated that explanatory feedback positively affected both learning gains and the learning process, while open-ended self-explanation did not. In a more recent study, Adams and Clark (2014) also investigated the effects of self-explanation prompts and feedback, but they employed *menu-based self-explanations*. They compared three conditions: (explanatory) feedback, menu-based self-explanations, and control (no explanatory feedback or self-explanation prompts). They found that there were no overall learning differences between the conditions, but condition influenced students' progress in the game, with students in the control condition completing significantly more levels in the game than the students in the self-explanation condition. Adams and Clark (2014) conjectured that the game they employed in their study was already highly interactive and therefore the intrinsic processing load was already high, making students unable to engage reflectively with the self-explanation prompts, and concluded that for students to be able to take advantage of explanation functionality tools designed to increase deeper understanding, it is important to manage cognitive load and game flow.

In a recent study, ter Vrugte et al. (2015) investigated the effects of menu-based self-explanations ('structured reflection prompts' in their study) and procedural information (i.e., explanatory feedback addressing general rules about the educational domain). In their study, students from prevocational education worked with an educational mathematics game named 'Zeldenrust' (Vandercruysse et al., 2015) which was designed to develop students' proportional reasoning. Students received either procedural information (visual representations of domain-related procedures/strategies), directive directive self-explanation prompts, a combination of procedural information and menu-based self-explanation prompts, or no support during the game. In this study, ter Vrugte et al. (2015) assumed that the self-explanation prompts would affect learning, and the procedural information could help the students to structure and represent their explanations, which would promote the acquisition of explicit knowledge while minimizing processing demands. However, the study revealed no effect of the menu-based self-explanations or the procedural information on game-based learning, and ter Vrugte, et al. (2015) consider whether the process of self-explanation was too demanding for the students involved. In addition, they conjecture that the timing and quantity of the prompts could have affected the results. They suggest that a more continuous and adaptive form of support would benefit the students more. A similar study was performed by Killingsworth et al. (2015), who compared the effects of menu-based self-explanation prompts and explanatory feedback explaining general rules of the targeted educational domain (comparable to the procedural information condition in the study by ter Vrugte et al. (2015)). Killingsworth et al. (2015) found no overall differences in learning gain between conditions, but found that when the number of levels completed was taken into account, students in the self-explanation condition showed higher learning gains than the students in the explanatory feedback condition, indicating that specific learner characteristics might influence the effectiveness of the prompts.

8.3.4 Conclusion

The results of studies indicate that the addition of self-explanation prompts does not guarantee positive effects. Most studies found no added value (Adams & Clark, 2014; Hsu & Tsai, 2011, 2013; Moreno & Mayer, 2005; ter Vrugte et al., 2015). Johnson and Mayer (2010) found added value, but only for directive prompts and they only evaluated transfer of knowledge within the game-context. Other studies only found added value when other constraints (accuracy of responses, and processing demands of self-explanations) were taken into account (Hsu et al., 2012; Killingsworth et al., 2015). From the literature to date, the conclusion can be drawn that the success of self-explanation prompts depends on their content and the way they are configured, and that their added value may depend on specific game features (e.g., interactivity). Most researchers highlight the fact that all prompts should be carefully tailored to fit students' characteristics, gameplay behavior, and strategies. In light of that consideration and previous results, it would be valuable to evaluate other means to elicit self-explanations in game-based learning.

8.4 Other Ways to Stimulate Self-Explanations?

8.4.1 Self-Explanation Through Collaboration

Instead of trying to get students to explain their thoughts to themselves (self-explanation), we could also trigger the same process by having them explain their thoughts to someone else. Any situation where students must collaborate can induce these kinds of explanations. In collaboration, explanations can occur in different forms, all fostering the generation of explicit knowledge. When students ask questions, they outline what they know and/or identify what they need to know, which is likely to help them generate explicit knowledge structures/schemas. Students who answer questions are made aware of their knowledge, and must consciously revisit their actions and verbalize what they know. Beyond simply prompting students to self-explain, collaboration can induce discussion. Discussion about conflicting information can complete existing mental models or induce their reconstruction, and the quality of the knowledge might improve correspondingly. In the context of game-based learning, collaboration allows for conscious reflection on the educational content and the construction of formal and explicit knowledge (Chen & Law, 2016).

Collaborative learning is a well-defined and thoroughly explored domain. Collaboration can be defined as a situation in which two or more people engage in problem solving and co-construct knowledge. A considerable number of studies show that both learning processes and learning outcomes can benefit from collaboration (e.g., Cohen, 1994; Kyndt et al., 2013; Webb, 1982). In addition, interaction in a collaborative setting is a highly engaging activity. In their ICAP framework, Chi and Wylie (2014) categorize different types of engagement activities and list interaction as most engaging. Because engagement is thought to positively affect learning, collaboration is likely to benefit learning more than less engaging activities. Although computers have been used to script and structure both face-to-face collaborative tasks and distance collaboration, research focusing on the effects of collaboration in game-based learning is relatively scarce. Some studies have discussed the effects of collaboration, but only a few studies have actually investigated the self-explanations that occur during these collaborations and how they affect learning.

When considering collaboration in game-based learning, we need to bear in mind that there is a difference between 'technology that is used collaboratively' (in which the software does not entail collaboration), and 'collaborative technology' (software designed to support collaboration). In game-based learning, this means that we should differentiate between games designed to play collaboratively (e.g., multiplayer games) and games that are simply played in a collaborative setting. Though we expect that the processes triggered by games played in a collaborative setting (questioning, answering, and discussing) are similar to those that would be elicited by collaborative technology such as multiplayer games, for the current overview we focus on the former.

An early study by Inkpen, Booth, Klawe, and Upitis (1995) reported a comparison of gameplay in individual, parallel, or collaborative settings. Students in the individual condition played alone, those in the parallel condition played side by side on their own computers, and those in the collaborative condition sat side by side and played together at the same computer. Inkpen et al. (1995) found that students in the collaborative condition had more verbal interactions than those in the parallel condition, and they performed better during the game than the students in the parallel and individual conditions. In addition, collaboration affected the motivation to continue playing. However, other studies have found no positive effect of collaboration on learning from educational games. A study by ter Vrugte et al. (2015) investigated how collaboration and competition would affect knowledge acquisition from an educational math game called 'Zeldenrust' (Vandercruysse et al., 2015). They had a 2×2 design with two factors: in-class competition and heterogeneous, face-to-face collaboration. Dyads were created based on the level of prior domain knowledge: grouping above average students with below average students. The collaboration was designed in accordance with the Student Teams Achievement Division model. Students were competing with their team score (a product of individual progress and team effort during the game) when they were in a collaborative and competitive condition, or with their individual score (a product of individual progress and individual effort during the game) when they were in a competitive and noncollaborative condition. ter Vrugte et al. (2015) found no overall effects of collaboration or competition, but found that students' level of prior knowledge influenced the interaction between collaboration and competition, suggesting that specific learner characteristics should be taken into account when designing for and evaluating the effects of collaboration.

Studies by van der Meij, Albers, and Leemkuil (2011) and Meluso, Zheng, Spires, and Lester (2012) investigated the effects on learning and engagement of playing a game in collaborative pairs or individually. Both studies found no differences between conditions. Meluso et al. (2012) conjecture that it might be that the actions of the players lacked specificity (actions were not always directed toward learning) in the collaborative condition, and van der Meij et al. (2011) argue that the collaboration might not have been successful due to a lack of depth in the dialogues: partners mainly discussed game features. Both studies recommended exploring the possibilities of scripted collaboration. This is in line with research on collaborative learning in general, which stresses the fact that freely collaborating students who do not have consistent support from the teacher usually fail to engage in productive learning interactions. Unfortunately, while research on collaboration in educational games seems scarce, experimental research focusing on scripted collaboration in educational games seems altogether lacking. From a case study on collaboration in a scripted educational game environment, Hamalainen (2008) concluded that scripts can make collaborative game-based learning more effective, but that group-specific learning processes make it difficult to successfully script collaboration without over-scripting it (which would constrain students' own constructions and limit the full capacity of collaborative learning). In a brief exploratory review on scripting in game-based learning, Demetriadis, Tsiatsos, and Karakostas (2012, p. 154) concluded that "scripted collaboration should be considered as a major guiding framework for the pedagogy and the architecture of digital learning games," but they also conclude that research still needs to be done.

A more recent study by Chen, Wang, and Lin (2015) also compared collaborative and individual gameplay. They found no quantitative differences between conditions concerning motivation or learning outcomes. Qualitative analyses of a postgame interview revealed that students had diverse experiences of the collaboration. Some interviews revealed that students did have meaningful interactions and felt that the collaboration was supportive. However, other students reported that they did not appreciate the collaboration. Chen et al. (2015) concluded that certain personality traits can clash and lead to an unproductive group. Another recent and more elaborate study by Chen and Law (2016) investigated the effects of scaffolding on secondary students' individual and collaborative game-based learning. They compared four conditions: collaborative, individual, collaborative with scaffolds, and individual with scaffolds. The scaffolds were open-ended questions provided to the collaborating students after the game. The questions asked students to make explicit connections between the game world and disciplinary knowledge. Chen and Law (2016) hypothesize that the questions could serve to support the collaboration and would improve the quality of discussion. Results showed that both the questions and the collaboration positively affected learning. In addition, it was found that the questions strengthened the already positive effect of collaboration on learning. However, both the collaboration and the questions had a negative impact on motivation. In the collaborative and question conditions, feelings of competence, interest, and autonomy were lower than in the individual and no-question conditions.

In conclusion, we can state that collaboration is a promising method of optimizing game-based learning. Among other positive effects, collaboration allows students to provide and receive explanations. This verbalization can help the students transform implicit knowledge into explicit knowledge. However, in practice the effects of collaboration on knowledge acquisition in game-based learning are not unequivocal (Wouters & van Oostendorp, 2013). When collaboration is left completely to the students' devices, the interactions seem to fall short on quality (focus and depth) and quantity. Providing scripts or scaffolds could positively affect quality and quantity of the interactions and stimulate generation of explicit knowledge. In addition, two studies have indicated that discussion after the game (guided by questions) positively affects learning (Chen & Law, 2016; van der Meij et al., 2011). This indicates that (guided) debriefing could be a successful method of employing the strength of discussion to optimize game-based learning.

8.4.2 Self-Explanation Through Partial Worked Examples

Moreno and Mayer (2005) discovered that self-explanation prompts were effective in interactive environments when students were asked to self-explain program-provided solutions rather than their own solutions. Presenting students with solutions is a more controlled way to initiate self-explanations, in that it provides the possibility of controlling the information the student is reflecting on. One way to provide this information is by means of worked examples. Worked examples are step-by-step expert explanations of how to solve a problem (Anderson, Fincham, & Douglass, 1997). Research indicates that exposure to worked examples can be very effective for skill acquisition in well-structured domains such as mathematics (Anderson et al., 1997; Carroll, 1994). In addition, worked examples provide expert models and therefore can be used as prompts to guide students and increase their efficiency, feelings of competence, and success (Carroll, 1994; Cooper & Sweller, 1987; Tarmizi & Sweller, 1988). Research has shown that students who interact with worked examples learn more when they explain the examples to themselves: the self-explanation effect (Chi et al., 1989, 1994). Though most students are likely to engage in some form of spontaneous self-explanation, the quality of these explanations varies, and most students are likely to use inadequate self-explanation strategies (i.e., passive or superficial) while studying worked examples (Renkl, 1997). Atkinson, Derry, Renkl, and Wortham (2000) found that the structure of the worked example can encourage students to actively self-explain. In a follow-up article, Atkinson and Renkl (2007) suggested fading (gradually removing worked-out steps) as a possible means of inducing self-explanations. Other research on partial worked examples has endorsed the positive effects on learning (Richey & Nokes-Malach, 2013): students process these worked examples more actively (Atkinson & Renkl, 2007; Paas, 1992; van Merriënboer & de Croock, 1992), and they are more encouraged to participate in self-explanation (Renkl, Atkinson, & Große, 2004).

Though worked examples have proven to be successful in other fields such as simulation-based learning, worked examples are not that common in game-based learning. In a recent study, we opted to use fading (partial) worked examples to optimize game-based learning (ter Vrugte et al., in preparation). In this study, we incorporated two conditions that were identical in terms of embedded learning objectives (proportional reasoning) and learning material (the game environment) and differed on only one variable: the presence or absence of worked examples in the game. The worked examples were all partial and the worked out steps gradually faded while the students progressed through the game (prompting students to extend their self-explanations as they progressed). Results of this study show that students in the worked-example condition performed significantly better on a domain knowledge posttest than students from the control condition. In addition, students from the worked-example condition were more able to explicate their knowledge on the posttest, suggesting that the addition of the faded worked examples did indeed help students to verbalize their knowledge and generate more explicit knowledge structures.

In conclusion, we can state that the use of worked examples seems promising based on worked examples theory and the results from our latest study, but little research has targeted worked examples in games. It could be that game developers hold back on providing formal representations of the educational content in the game, because they fear that these might diminish students' motivation and experience of game flow. We feel that this fear might be unjust. Based on findings of O'Neil et al. (2014) and ter Vrugte et al. (in preparation), we speculate whether prompts that specifically elucidate the link between the formal educational content and the implicit educational content in the game are exactly what we need to optimize game-based learning.

8.5 General Implications and Guidelines

At the beginning of the chapter, we not only introduced game-based learning as a potential effective instructional approach, but also pointed out that the effects of game-based learning are varied and far from optimal. Aside from many features and characteristics that might affect the results of game-based learning, we conjectured that games generally thrive on experiential learning and experiential learning increases knowledge, but that this knowledge is often implicit. We noted that though implicit knowledge is certainly valuable, in general explicit knowledge is considered more desirable in education, because it is more accessible and promotes transfer. It is suggested that explicit knowledge does not always automatically follow from the development of implicit knowledge, but that this process can be supported through self-explanations. Because self-explanations rarely occur automatically in game-based learning environments, we proposed that self-explanations in game-based learning environments can be elicited by specific instructional approaches. Three possible approaches for eliciting self-explanations were discussed: question prompts, collaboration, and partial worked examples.

Although in general, the literature on self-explanations finds that these are a key aspect of formal learning, results of research on self-explanations in game-based learning seem less straightforward. Overall, we can conclude that the outcomes are varied. Because the majority of studies focused on outcome measures such as learning gains or motivation, the actual underlying process is generally inferred, but rarely investigated. Monitoring verbalizations in collaborative learning and think-aloud protocols when learning with (fading) worked examples or self-explanation prompts could provide more insight into the actual processes involved. Identifying how students process and interact with the scaffolds, and what learner and game characteristics affect/shape these processes can help scientists to optimize game-based learning. Despite these research design considerations, we would like to point out some noteworthy research findings from the studies discussed previously, and infer some general guidelines.

From studies that explored the impact of accuracy in self-explanations, we can deduce that an increase in the accuracy of self-explanations could foster learning, though it is not entirely clear how student's ability and prior knowledge are involved. Instead of trying to elicit accurate self-explanations, we could also provide students with accurate explanations. However, game-based learning research comparing explanatory feedback to self-explanations is also inconclusive. Providing students with explanatory feedback not only is likely to reduce the processing load and increases the accuracy of the information (explanations), but it also reduces students' engagement and chances for active processing.

Hsu et al. (Hsu & Tsai, 2013; Hsu et al., 2012, in press) defined students' engagement in terms of accuracy; however, though we agree that these two are likely to be correlated, we feel that low accuracy does not mean that the student was not engaged. It could be that the student did not possess the relevant knowledge to respond to the self-explanation question correctly. When considering how to categorize students' cognitive engagement in self-explaining, we would suggest the adoption of a categorization such as the ICAP framework suggested by Wylie and Chi (2014). In this framework, students' engagement is appraised based on their overt behavior. ICAP stands for interactive, constructive, active, and passive, with interactive being the most cognitively engaging and passive being the least cognitive engaging. The label 'passive' would fit the reception of explanatory feedback: students receive explanations and a minimal level of engagement is required. The label 'active' would fit supported self-explanation conditions (i.e., source-based, menu-based, and fill-in-the-blank) in which students must show some level of engagement in order to obtain the explanations. The label 'constructive' seems to fit best with open-ended and focused self-explanations, where students actively construct their self-explanations based on limited provision of material.

In addition, we want to call attention to the fact that most of the studies discussed provided no feedback on students' self-explanations, even though feedback could affect the quantity and quality of students' self-explanations (Aleven & Koedinger, 2002; Cheshire, Ball, & Lewis, 2005; Kwon, Kumalasari, & Howland, 2011). When students receive no (explanatory) feedback on their response, existing, incorrect mental models can persist, and this is likely to diminish the effect of self-explanations on learning. In line with this is a recommendation by Chi (2000), who proposes a reflection mechanism in self-explanations to draw students' attention to possible differences between their mental model and the learning material.

When students collaborate they can provide feedback on each other's explanations (discussion), so that discussion can serve as a reflective activity. However, results of studies that have addressed game-based learning and collaboration seem to suggest that this discussion also needs to be guided to be effective (e.g., by providing scripts or questions that can focus the discussion). In addition, it seems that reflection after action (a reflective activity after collaborative play, e.g., a debriefing) can further benefit collaborative game-based learning. In their study, van der Meij et al. (2011) had students answer questions collaboratively after the posttest. They deduced that this activity positively affected learning. A similar finding was demonstrated in the study by Chen and Law (2016), who tried to scaffold the collaboration by having students answer questions collaboratively after the game. They found that this strengthened the effect of the collaboration. These findings are in line with studies on debriefing in which retrieval processes are used to affect learning. It seems noteworthy to mention that the questions used by Chen and Law (2016) match the category of question prompts that O'Neil et al. (2014) had found to elicit the most effective self-explanations. Both findings are in line with the observation that it is important that students see the connection between what they do in the game and the formal learning content: both debriefing and focused prompts can help students make this connection. We feel that this finding can serve as a guideline for the design of instructional support for game-based learning.

Although the literature to date does not unanimously indicate that game-based learning environments with elicited self-explanations are more effective than game-based learning environments without them, we think eliciting self-explanations can be beneficial for game-based learning when certain design considerations are taken into account: *accuracy seems to matter*, students need *guidance* to focus their explanations, and processing demands should be minimized. Therefore, we suggest directive/focused prompts. Prompts are most likely to be effective when they *elicit self-explanations that clarify the relation between the game content and the educational content.* In addition, *feedback* can help to remedy incorrect explanations and can help repair mental models. Direct feedback, discussion, and debriefing are options for optimizing self-explanations in game-based learning. Partial worked examples seem to conform to the above-mentioned design considerations. Worked examples can offer a means of introducing formal representations of educational content in a game-based learning environment. In addition, when worked examples are incomplete, they are more likely to elicit self-explanations. The addition of feedback, discussion, or debriefing can further optimize results.

Source of Funding Sponsored by NWO under grant number 411-10-900 and FWO under grant number G.0.516.11.N.10.

References

Adams, D. M., & Clark, D. B. (2014). Integrating self-explanation functionality into a complex game environment: Keeping gaming in motion. *Computers & Education, 73*, 149–159. doi:10.1016/j.compedu.2014.01.002.

Aleven, V. A. W. M. M., & Koedinger, K. R. (2002). An effective metacognitive strategy: Learning by doing and explaining with a computer-based cognitive tutor. *Cognitive Science, 26*, 147–179. doi:10.1016/S0364-0213(02)00061-7.

Anderson, J. R., Fincham, J. M., & Douglass, S. (1997). The role of examples and rules in the acquisition of a cognitive skill. *Journal of Experimental Psychology: Learning, Memory, and Cognition, 23*, 932.

Atkinson, R. K., Derry, S. J., Renkl, A., & Wortham, D. (2000). Learning from examples: Instructional principles from the worked examples research. *Review of Educational Research, 70*, 181–214. doi:10.3102/00346543070002181.

Atkinson, R. K., & Renkl, A. (2007). Interactive example-based learning environments: Using interactive elements to encourage effective processing of worked examples. *Educational Psychology Review, 19*, 375–386. doi:10.1007/s10648-007-9055-2.

Barab, S. A., Thomas, M., Dodge, T., Carteaux, R., & Tuzun, H. (2005). Making learning fun: Quest atlantis, a game without guns. *Educational Technology Research and Development, 53*, 86–107. doi:10.1007/bf02504859.

Barzilai, S., & Blau, I. (2014). Scaffolding game-based learning: Impact on learning achievements, perceived learning, and game experiences. *Computers & Education, 70*, 65–79. doi:10.1016/j.compedu.2013.08.003.

Berry, D. C., & Broadbent, D. E. (1984). On the relationship between task performance and associated verbalizable knowledge. *The Quarterly Journal of Experimental Psychology Section A, 36*, 209–231. doi:10.1080/14640748408402156.

Berthold, K., Eysink, T. H. S., & Renkl, A. (2009). Assisting self-explanation prompts are more effective than open prompts when learning with multiple representations. *Instructional Science, 37*, 345–363. doi:10.1007/s11251-008-9051-z.

Boud, D., Keogh, R., & Walker, D. (1985). *Reflection: Turning experience into learning*. London: Routledge & Kegan Paul.

Carroll, W. M. (1994). Using worked examples as an instructional support in the algebra classroom. *Journal of Educational Psychology, 86*, 360–367. doi:10.1037/0022-0663.86.3.360.

Chen, C., & Law, V. (2016). Scaffolding individual and collaborative game-based learning in learning performance and intrinsic motivation. *Computers in Human Behavior, Part B, 55*, 1201–1212. doi:10.1016/j.chb.2015.03.010.

Chen, C., Wang, K., & Lin, Y. (2015). The comparison of solitary and collaborative modes of game-based learning on students' science learning and motivation. *Journal of Educational Technology & Society, 18*, 237–248. doi:10.2307/jeductechsoci.18.2.237.

Cheshire, A., Ball, L. J., & Lewis, C. N. (2005). In B. G. Bara, L. Barsalou, & M. Bucciarelli (Eds.), *Self-explanation, feedback and the development of analogical reasoning skills: Microgenetic evidence for a metacognitive processing account*. Paper presented at the Proceedings of the Twenty-Second Annual Conference of the Cognitive Science Society.

Chi, M. T. H. (2000). Self-explaining expository texts: The dual processes of generating inferences and repairing mental models. In R. Glaser (Ed.), *Advances in instructional psychology* (pp. 161–238). Mahwah, NJ: Lawrence Erlbaum Associates.

Chi, M. T. H., Bassok, M., Lewis, M. W., Reimann, P., & Glaser, R. (1989). Self-explanations: How students study and use examples in learning to solve problems. *Cognitive Science, 13*, 145–182. doi:10.1016/0364-0213(89)90002-5.

Chi, M. T. H., De Leeuw, N., Chiu, M., & Lavancher, C. (1994). Eliciting self-explanations improves understanding. *Cognitive Science, 18*, 439–477. doi:10.1016/0364-0213(94)90016-7.

Chi, M. T. H., & Wylie, R. (2014). The icap framework: Linking cognitive engagement to active learning outcomes. *Educational Psychologist, 49*, 219–243. doi:10.1080/00461520.2014.965823.

Cohen, E. G. (1994). Restructuring the classroom: Conditions for productive small groups. *Review of Educational Research, 64*, 1–35. doi:10.2307/1170744.

Cooper, G., & Sweller, J. (1987). Effects of schema acquisition and rule automation on mathematical problem-solving transfer. *Journal of Educational Psychology, 79*, 347–362. doi:10.1037/0022-0663.79.4.347.

Demetriadis, S., Tsiatsos, T., & Karakostas, A. (2012). *Scripted collaboration to guide the pedagogy and architecture of digital learning games*. Paper presented at the Proceedings of the European Conference on Games Based Learning.

Eraut, M. (2000). Non-formal learning and tacit knowledge in professional work. *British Journal of Educational Psychology, 70*, 113–136. doi:10.1348/000709900158001.

Girard, C., Ecalle, J., & Magnan, A. (2013). Serious games as new educational tools: How effective are they? A meta-analysis of recent studies. *Journal of Computer Assisted Learning, 29*, 207–219. doi:10.1111/j.1365-2729.2012.00489.x.

Hamalainen, R. (2008). Designing and evaluating collaboration in a virtual game environment for vocational learning. *Computers & Education, 50*, 98–109. doi:10.1016/j.compedu.2006.04.001.

Hsu, C. Y., & Tsai, C. C. (2011). Investigating the impact of integrating self-explanation into an educational game: A pilot study. In M. Chang, W.-Y. Hwang, M.-P. Chen, & W. Müller (Eds.),

Edutainment technologies. Educational games and virtual reality/augmented reality applications (Vol. 6872, pp. 250–254). Berlin: Springer.

Hsu, C. Y., & Tsai, C. C. (2013). Examining the effects of combining self-explanation principles with an educational game on learning science concepts. *Interactive Learning Environments, 21*, 104–115. doi:10.1080/10494820.2012.705850.

Hsu, C. Y., Tsai, C. C., & Wang, H. Y. (2012). Facilitating third graders' acquisition of scientific concepts through digital game-based learning: The effects of self-explanation principles. *The Asia Pacific Education Researcher, 21*, 71–82.

Hsu, C. Y., Tsai, C. C., & Wang, H. Y. (2016). Exploring the effects of integrating self-explanation into a multi-user game on the acquisition of scientific concepts, *Interactive Learning Environments, 24*, 844–858. doi:10.1080/10494820.2014.926276.

Inkpen, K., Booth, S. K., Klawe, M., & Upitis, R. (1995). *Playing together beats playing apart, especially for girls*. Paper presented at the First International Conference on Computer Support for Collaborative Learning, Indiana University, Bloomington, Indiana, USA.

Johnson, C. I., & Mayer, R. E. (2010). Applying the self-explanation principle to multimedia learning in a computer-based game-like environment. *Computers in Human Behavior, 26*, 1246–1252. doi:10.1016/j.chb.2010.03.025.

Jordi, R. (2010). Reframing the concept of reflection: Consciousness, experiential learning, and reflective learning practices. *Adult Education Quarterly, 61*, 181–197. doi:10.1177/0741713610380439.

Ke, F. (2008). A case study of computer gaming for math: Engaged learning from gameplay? *Computers & Education, 51*, 1609–1620. doi:10.1016/j.compedu.2008.03.003.

Kiili, K. (2005). Digital game-based learning: Towards an experiential gaming model. *The Internet and Higher Education, 8*, 13–24. doi:10.1016/j.iheduc.2004.12.001.

Killingsworth, S. S., Clark, D. B., & Adams, D. M. (2015). Self-explanation and explanatory feedback in games: Individual differences, gameplay, and learning. *International Journal of Education in Mathematics, Science and Technology, 3*, 162–186.

Kwon, K., Kumalasari, C. D., & Howland, J. L. (2011). Self-explanation prompts on problem-solving performance in an interactive learning environment. *Journal of Interactive Online Learning, 10*, 96–112.

Kyndt, E., Raes, E., Lismont, B., Timmers, F., Cascallar, E., & Dochy, F. (2013). A meta-analysis of the effects of face-to-face cooperative learning. Do recent studies falsify or verify earlier findings? *Educational Research Review, 10*, 133–149. doi:10.1016/j.edurev.2013.02.002.

Leemkuil, H., & de Jong, T. (2012). Adaptive advice in learning with a computer based strategy game. *Academy of Management Learning and Education, 11*, 653–665. doi:10.5465/amle.2010.0141.

Mayer, R. E. (2005). Cognitive theory of multimedia learning. In R. E. Mayer (Ed.), *The Cambridge handbook of multimedia learning* (pp. 31–48). New York: Cambridge University Press.

Meluso, A., Zheng, M., Spires, H. A., & Lester, J. (2012). Enhancing 5th graders' science content knowledge and self-efficacy through game-based learning. *Computers & Education, 59*, 497–504. doi:10.1016/j.compedu.2011.12.019.

Moreno, R., & Mayer, R. E. (2005). Role of guidance, reflection, and interactivity in an agent-based multimedia game. *Journal of Educational Psychology, 97*, 117–128. doi:10.1037/0022-0663.97.1.117.

O'Neil, H. F., Chung, G. K. W. K., Kerr, D., Vendlinski, T. P., Buschang, R. E., & Mayer, R. E. (2014). Adding self-explanation prompts to an educational computer game. *Computers in Human Behavior, 30*, 23–28. doi:10.1016/j.chb.2013.07.025.

Paas, F. G. (1992). Training strategies for attaining transfer of problem-solving skill in statistics: A cognitive-load approach. *Journal of Educational Psychology, 84*, 429–434. doi:10.1037/0022-0663.84.4.429.

Reber, S. (1993). *Implicit learning and tacit knowledge: An essay on the cognitive unconscious*. New York: Clarendon.

Renkl, A. (1997). Learning from worked-out examples: A study on individual differences. *Cognitive Science, 21*, 1–29. doi:10.1207/s15516709cog2101_1.

Renkl, A. (2002). Worked-out examples: Instructional explanations support learning by self-explanations. *Learning and Instruction, 12*, 529–556. doi:10.1016/S0959-4752(01)00030-5.

Renkl, A., Atkinson, R. K., & Große, C. S. (2004). How fading worked solution steps works—a cognitive load perspective. *Instructional Science, 32*, 59–82. doi:10.1023/B:TRUC.0000021815.74806.f6.

Richey, J. E., & Nokes-Malach, T. J. (2013). How much is too much? Learning and motivation effects of adding instructional explanations to worked examples. *Learning and Instruction, 25*, 104–124. doi:10.1016/j.learninstruc.2012.11.006.

Roy, M., & Chi, M. T. H. (Eds.). (2005). *The self-explanation principle in multimedia learning*. New York: Cambridge University Press.

Sun, R., Merrill, E., & Peterson, T. (2001). From implicit skills to explicit knowledge: A bottom-up model of skill learning. *Cognitive Science, 25*, 203–244. doi:10.1016/S0364-0213(01)00035-0.

Tarmizi, R. A., & Sweller, J. (1988). Guidance during mathematical problem solving. *Journal of Educational Psychology, 80*, 424–436. doi:10.1037/0022-0663.80.4.424.

ter Vrugte, J., & de Jong, T. (2012). How to adapt games for learning: The potential role of instructional support. In S. Wannemacker, S. Vandercruysse, & G. Clarebout (Eds.), *Serious games: The challenge* (Vol. 280, pp. 1–5). Berlin: Springer.

ter Vrugte, J., de Jong, T., Vandercruysse, S., Wouters, P., van Oostendorp, H., & Elen, J. (2015). How competition and heterogeneous collaboration interact in prevocational game-based mathematics education. *Computers & Education, 89*, 42–52. doi:10.1016/j.compedu.2015.08.010.

ter Vrugte, J., de Jong, T., Vandercruysse, S., Wouters, P., van Oostendorp, H., & Elen, J. (in preparation). Game based mathematics education: Do fading worked examples facilitate knowledge acquisition?

ter Vrugte, J., de Jong, T., Wouters, P., Vandercruysse, S., Elen, J., & van Oostendorp, H. (2015). When a game supports prevocational math education but integrated reflection does not. *Journal of Computer Assisted Learning, 31*, 462–480. doi:10.1111/jcal.12104.

van der Meij, H., Albers, E., & Leemkuil, H. (2011). Learning from games: Does collaboration help? *British Journal of Educational Technology, 42*, 655–664. doi:10.1111/j.1467-8535.2010.01067.x.

van Merriënboer, J. J. G., & de Croock, M. B. M. (1992). Strategies for computer-based programming instruction: Program completion vs. Program generation. *Journal of Educational Computing Research, 8*, 365–394. doi:10.2190/mjdx-9pp4-kfmt-09pm.

Vandercruysse, S., ter Vrugte, J., de Jong, T., Wouters, P., van Oostendorp, H., Verschaffel, L., et al. (2015). "Zeldenrust": A mathematical game-based learning environment for prevocational students. In J. Torbeyns, E. Lehtinen, & J. Elen (Eds.), *Describing and studying domain-specific serious games* (pp. 63–81). Cham: Springer International Publishing.

Webb, N. M. (1982). Peer interaction and learning in cooperative small groups. *Journal of Educational Psychology, 74*, 642–655. doi:10.1037/0022-0663.74.5.642.

Wouters, P., Paas, F., & van Merriënboer, J. G. (2008). How to optimize learning from animated models: A review of guidelines based on cognitive load. *Review of Educational Research, 78*, 645–675. doi:10.3102/0034654308320320.

Wouters, P., & van Oostendorp, H. (2013). A meta-analytic review of the role of instructional support in game-based learning. *Computers & Education, 60*, 412–425. doi:10.1016/j.compedu.2012.07.018.

Wylie, R., & Chi, M. T. H. (2014). The self-explanation principle in multimedia learning. In R. E. Mayer (Ed.), *The Cambridge handbook of multimedia learning* (2nd ed., pp. 413–432). New York: Cambridge University Press.

Chapter 9
Competition and Collaboration for Game-Based Learning: A Case Study

Eric Sanchez

Abstract This chapter is dedicated to discuss how competition and collaboration for playing have an impact on the learner's strategy and the learning process. We examine what kind of effects can be expected when competition and collaboration are used as instructional techniques. This is done through a discussion about how these concepts are defined and through a brief overview of the literature dedicated to competition and collaboration in game-based learning contexts. We also discuss the results obtained with an empirical study carried out with *Tamagocours*, an online multiplayer Tamagotchi-like game dedicated to teach preservice teachers how to follow the legal rules that should be applied for the use of digital resources in educational contexts. In this chapter, we consider that playing encompasses two dimensions: an individual and conflictual play with an antagonist system and a collaborative play which is based on cooperation with the game (to accept to play) and/or with teammates. Game-based learning is thus *coopetitive* and results from conflictual interactions and *epistemic* interactions when players collaborate.

Keywords Game-based learning • Competition • Collaboration

9.1 Introduction

Berlin, 31st of October 2015, Christopher Seitz is participating to the final step of *League of Legend* world championship. *LoL* is one of the nowadays most played third-person multiplayer online battle arena (MOBA) games. Christopher stays immobile in front of its screen, and rapidly clicks on his mouse and presses the keys of his keyboard. He is known as one of the best players of the world and he is able to click at a pace close to 400 times per minute. He is concentrated on one objective:

E. Sanchez (✉)
Centre d'enseignement et de recherche pour la formation a l'enseigment au secondaire I et II, Université de Fribourg, Rue P.-A. de Faucigny 2, Fribourg 1700, Swiss
e-mail: eric.sanchez@unifr.ch

P. Wouters, H. van Oostendorp (eds.), *Instructional Techniques to Facilitate Learning and Motivation of Serious Games*, Advances in Game-Based Learning, DOI 10.1007/978-3-319-39298-1_9

destroying the opposing team's "nexus," a building that lies at the heart of a base protected by defensive structures. In order to achieve this, different actions must be performed: facing the monsters of the "jungle," killing other champions, destroying defensive turrets, and dealing with the final blow to enemy mines. Christopher is also talking in his headset. Indeed, he is a player of *SK gaming*, a German team involved in the video games world championship since more than five years and the game outcome will depend on the capacity of the team members to collaborate. As a single-player Christopher competes to achieve the best score and to get a better ranking for his own. He faces computer online generated opponents or adversaries from other teams. He also collaborates with his teammates in order to find the best strategy to achieve the game.

Many games, like *League of Legend*, are a subtle combination of competition and collaboration, of competition with opponents automatically generated by computer or avatars driven by adversaries and collaboration with face-to-face or online teammates. This combination is not a specific feature of modern online video games and the well-known *prisoner's dilemma* formalized in 1950 by Albert W. Tucker (1905–1995) is also a good illustration of a game in which the outcome depends on the balance between cooperation and competition and, in this case, of two purely "rational" individuals who do not cooperate, even if it appears that it is in their best interests to do so.

Competition and collaboration are also known to be both in strong connection with learning (Burguillo, 2010; Johnson & Johnson, 1996). On the one hand, educational systems are mainly driven by marks, grades, and competitive exams. Therefore, competition is a core element of learning in formal contexts such as schools or universities. On the other hand, constructivist theory of learning has now demonstrated the importance of collaborative learning and educators try to design learning settings based on the learners' collaboration (Dillenbourg, 1999). There is an increasing usage of games for educational purposes that address the issue of competition and collaboration. As a result, we consider that studies on game-based learning are relevant both for theoretical purposes (we need to understand how collaboration and competition influence learning with games) and for pragmatic purposes (we need to develop guidelines for game designers and educators).

This chapter is dedicated to discuss how competition and collaboration in a game have an impact on the learning process. We will examine what kind of effects can be expected when competition and collaboration are used as instructional techniques. This will be done through an brief overview of the literature dedicated to competition and collaboration in game-based learning contexts and also through the discussion of the results obtained with an empirical study carried out with *Tamagocours*, an online multiplayer Tamagotchi-like game dedicated to teach preservice teachers how to follow the legal rules that should be applied for the use of digital educational resources.

In the first section, we briefly introduce competition and collaboration and summarize how these concepts are defined and linked with game-based learning. The second section is dedicated to describe the methodology of an empirical study carried out with *Tamagocours*. In the three next sections, we discuss the lessons learned from this study: how competition and collaboration are linked to the commitment of the player with the game, how competition is in connection with learning through experience, and how collaboration enables the formulation of knowledge.

9.2 Toward the Concepts: Collaboration and Competition

9.2.1 Games and Competition

In his famous book, *Homo Ludens* published in 1944, Huizinga points out that though the ancient Greeks made a distinction between *âgon* (competition and contest) and *paidià* (spontaneous play of children), there are close connections between play and competition. The main thesis of Huizinga is that due to an innate universal disposition to *competition*, play has a central role in human culture. Play contests establish a social order that is merely symbolic, but the meaning of win or loss is interpreted concretely within the culture. He concludes that play lies at the beginning of each competition. Caillois, a French sociologist, disputes Huizinga's emphasis on competition. His classification system of play (1958) consists of four categories: *Alea* (games based on chance), *Mimicry* (games based on make-believe), *Ilinx* (games based on exhilaration), and *Agôn* (games based on competition). For Caillois competition consists of an artificial fight with equal opportunities and ideal conditions so that the victory of the winner will have an incontestable value.

Following the seminal work of Huizinga and Callois, the definitions for game proposed by many authors encompass the idea of competition. For example, according to Parlett (1999) a game is a contest to achieve an objective. Similarly, Abt (1970) points out that a game is a contest. He also emphasizes the importance of the rules among adversaries trying to win objectives. Suits (1990) considers that playing a game is a voluntary effort to overcome "unnecessary obstacles" (p. 34), and the computer game designer, Chris Crawford, uses the terms conflict, challenge, and opponent to describe what is a game (Crawford, 1982). The terms contest, conflict, challenge, adversary, and obstacle are widely used and the semantic field usually associated to the idea of play is similar to the semantic field associated to the idea of competition. As a result, Salen and Zimmerman (2004) conclude that all games are competitive. Competition is indeed a core element of games. Meaning can stem from the joy of play itself, but meaning also derives from the competitive struggle of a game (Salen & Zimmerman, 2004).

The "game" Progress Wars[1] (Stjerning, 2010) has been designed to demonstrate that playfulness is strongly connected with competition. This online "game" is a caricature of what is a game. The goal of Progress Wars consists of filling-in a progress bar by clicking the "perform mission" button (Fig. 9.1). There is no real challenge to address, no difficulty to deal with. After some clicks, the level is considered to be completed and a new empty progress bar appears. Then, a new "mission"

[1] http://www.progresswars.com

Fig. 9.1 Game Progress Wars 1 (Stjerning, 2010)

starts and the "player" is asked to click again in order to complete a next level, etc. Due to the absence of real challenge and competition for achieving the goal of the "game," the playfulness is clearly missing and playing this game does not make any sense. This example makes clear that the pleasure that results from playing is strongly linked to the capacity of the player to reach difficult but achievable objectives provided by competition.

A definition for the concept of competition is offered by game theory: "the zero-sum game. In zero-sum games one player's winning equals another player's losses. If one player is the victor in a two-player zero-sum game, the other player will necessarily lose. Winning is always equally balanced by losing, making the end sum zero" (Salen & Zimmerman, 2004, p. 255). However, amongst the diversity of games, competition takes different forms. A common element of video games is a competitive mode. Players compete with one another. This competition can also take another form. Two or more players compete for the same goal (Plass et al., 2013). As a result, there are two categories of competition: the adversary can be a human being as opponent or the game itself (Alessi & Trollip, 2001). Digital technologies offer the opportunity to develop games that can take the role of automatic opponents. For example, for chess games that have been traditionally played between humans, there are now numerous versions of electronic chessboards, personal computer software, or online chess games that allow a single player to compete with a computer program as an automated adversary. In other games such as MOBA (Multiplayer Online Battle Arena) games, puzzle games, or platform games, the challenges represent a resistance to the actions that the player performs to reach his goal. Therefore, playing a game also means competing against a game to overcome automated obstacles.

It is also worthwhile to notice that the context of a game may change the meaning of competition (Alessi & Trollip, 2001). For example, when *League of Legend* is played during a world championship, the stakes of the competition (rewards that consist of US$1,000,000 champion prize) dramatically change the consequences of a victory and amusement is then probably not the main objective for playing. In addition, a game does not generally consist of a unique competition but of a series of challenges, with increasing difficulties, at the different game levels. Each level consists of a competition by itself and a victory when the game is completed. For platform games, the final combat often consists of fighting with the "boss," the most powerful enemy of the game.

9.2.2 Games and Conflict

Competition is a core element of a game; however, the term *conflict* is often preferred to the term competition when it comes to propose a definition for games. Conflict has a larger scope than competition as it explicitly involves both competition with opponents and competition with the game system. Thus, conflict is a core element in the definition of game proposed by Salen and Zimmerman (2004):

> A game is a system in which players engage in an artificial conflict, defined by rules, that result in a quantifiable outcome.

In this respect, this definition follows the ideas already expressed by Crawford (1982):

> Conflict arises naturally from the interaction in a game. The player is actively pursuing some goal. Obstacles prevent him from easily achieving this goal. Conflict is an intrinsic element of all games. It can be direct or indirect, violent or nonviolent but it is always present in every game (p. 249)

This conflict arises in the *magic circle*, a metaphor coined by Huizinga (1955) to articulate the spatial, temporal, and psychological boundaries between games and the "real world." For Huizinga, play is a *magic circle*, a space separated from the outside world and defined by the rules enacted within its boundaries. The metaphor of the *magic circle* has been applied to the experiential dimension of gameplay which involves entering a particular experiential mode. This magic circle imbues games with special meanings and the game's victory conditions are one of the most important emerging meanings (Salen & Zimmerman, 2004). As a result, action and winning the game might only have value within the *magic circle* and the narrative frame of the game (Salen & Zimmerman, 2004). The notion of conflict entails goals to pursue (Juul, 2003). From a game design perspective, this can lead to an *economic approach* of the availability of resources for the player. For example, Schlieder, Kiefer, and Matyas (2006) propose a general solution for location-based games design. For location-based games, the gameplay depends on the player's location represented by points of interest (POIs). The number and the location of the POIs are chosen in a way that they maintain the conflict. During the game, POIs remain available and, for a given player, there is always something to "win."

Within a game, a conflict, as an intrinsic element, encompasses a contest of powers and/or a struggle for limited resources which leads to competition. Conflict entails different categories. For example, in a location-based game conflict arises for territory and video games, and gameplay often consists of a conflict for resources materialized by points, rewards, and leaderboards. The games that are based on the ability of the player to answer questions consist of a conflict over knowledge. This conflict can take different forms: an individual or a group of players can compete collaboratively or in parallel against an individual player, a team of players, or a game system. As a consequence, the adversaries or/and the game system form an *antagonist system* which resists to the actions of the player. Hence, it demands from the player to shape strategies and to develop the knowledge needed to win. In next sections, we discuss how a conflict with this *antagonist system* formed by the game (including the competitors) is a core element for the learning process in game-based learning.

9.2.3 Games, Cooperation, and Collaboration

Since more than 30 years, a lot of work has been carried out in the field of computer-supported collaborative learning (CSCL); however, Lipponen (2002) emphasizes that there is still no unifying and established theoretical framework. Objects of study are still not agreed upon and no methodological consensus or agreement about the concept of collaboration is reached. There are differences between collaboration and cooperation. Cooperative work is accomplished by the division of labor among participants. For cooperative work, each person is responsible for a portion of the problem solving. Collaboration involves mutual engagement of participants and coordinated efforts so that the problem will be solved (Roschelle & Teasley, 1995). The difference between collaboration and cooperation lies in the way labor is carried out. However, the two words are often used as synonyms.

Game theory shows that collaboration can come in a variety of formats. Obviously collaboration can occur not only between teammates who share a common goal but also between a player and his opponent. The *prisoners' dilemma* illustrates this idea. For this game, the calculation of probability demonstrates that the best strategy consists of defection or cooperation depending on whether the game is played in one tournament (i.e., a single and short time competition) or iteratively, taking also into account how the opponent reacts.

Collaboration plays an important role in the constructivist theories that emphasize the role of social interactions for the learning process. For example, in his seminal work Wenger (1998) states that learning consists of social participation. The individual, as an active participant in a *community of practice*, constructs his identity through this participation and the interactions that emerge within the community. *Epistemic interactions* are explanatory and argumentative interactions that play a role in the co-construction of scientific knowledge (Ohlsson, 1995). They can involve different interactive processes such as explanation, production of an articulated discourse, elaboration of meaning, or clarification of views (Baker, 1999). As a result, epistemic interactions play an important role in game-based learning and, in previous studies (Sanchez et al., 2012; Sanchez, Emin Martinez, & Mandran, 2015), we have proposed to avoid the widely used expression *serious game* that focuses on the artifact used to play and to prefer the expression *digital epistemic play* to name playful situations designed with digital technologies that intend to foster *epistemic interactions*.

For game-based learning, cooperation takes place when a player accepts to enter the *magic circle* and agrees to take on the arbitrary rules of the game. As a result, to play a game consists of cooperatively taking on its artificial meaning (Salen & Zimmerman, 2004) and the *lusory attitude* can be interpreted as the acceptation by the player to cooperate so that play can emerge. It means that in a way, all games are cooperative (Salen & Zimmerman, 2004) and the space of the game is cooperatively formed by the players in order to create competition and amusement (Salen & Zimmerman, 2004).

9.2.4 Collaboration and Competition for Game-Based Learning

In this section, we argue that though research provides evidence for the positive influence of competition and collaboration on attitudes toward learning, there is often an ambiguity on how they influence the learning process. In addition, we argue that there are two main ways for considering the influence of competition and collaboration: Play is merely considered to be an instructional approach that consists of fooling students by hiding educational objectives under game-like techniques (impact on students' motivation) or play is considered to have an intrinsic educational value. Subsequently, I we discuss models for game-based learning that try to take advantage from the positive effect of a combination of collaboration and competition. Finally, I we provide explanations for negative impacts that have also been reported.

Games appear to be a complex combination of cooperation and conflict: "The effort each player makes to overcome the resistance and achieve the goal is the heart of the game and what makes it enjoyable and gratifying. In most games, the resistance is supplied by your opponent trying to achieve her goal. Your opponent is therefore your partner in the game" (Fluegelman & Tembeck, 1976, pp. 87–88). This dimension has been early recognized by Huizinga (1955) who considered that a game creates social groups that separate themselves from the outside world. Thus, play is a form of contest that involves both cooperation (in agreeing to rules) and competition (so that it is possible to identify winners and losers). In the 1970s, a group of game designers, the New Games Movement, challenged conventional notions of games as conflict and affirmed the interdependent relationship between competition and cooperation (Salen & Zimmerman, 2004). Therefore, from an educational point of view, it might not be relevant to oppose competition and collaboration but merely to consider that they are two interdependent and inseparable core dimensions. In the following, we mention studies that have been dedicated to assess their influence on the learning process and discuss the explanations that have been proposed by different authors.

The effects of collaboration and competition on the learning process have been assessed by a considerable number of studies. Generally, the use of technology for collaboration is considered to be more effective than competition in increasing achievement and promoting positive attitudes and cognitive development (Johnson & Johnson, 1996). Though most of these studies do not involve games, there is evidence to suggest that collaboration with other players can positively impact learning gains. Ke (2008) reports the findings of a study on educational computer games used within various classroom situations. This study establishes the positive effect of "cooperative goal structures," as opposed to "competitive goal structures" on students' attitudes toward math learning but failed to demonstrate a significant impact on cognitive math test performance. Inkpen, Booth, Klawe, and Upitis (1995) investigated the "grouping effect" of a puzzle game called *The Incredible Machine*. They found that Female/Female pairs playing together on a machine reached significantly higher levels of performance than Female/Female pairs playing side by side on two computers. They also found that the level of motivation to continue playing the game was positively affected by the opportunity to collaborate with a partner. However, not all the studies that investigated the effects of collaboration provide clear results. For example Meluso, Zheng, Spires, and Lester (2012) found no significant difference on science learning between collaborative gameplay and single game player conditions, though the same authors report that when conditions were combined, science content learning increased. All in all, competition and collaboration in games are generally recognized to have a positive effect on attitudes toward learning. However, there is often an ambiguity on *how* competition and collaboration within a game-based context influence the learning outcomes and research provides limited empirical evidence on this aspect.

Some studies are conducted within a humanist tradition of teaching based on Erasmus (1467–1536). Play is here often considered to be merely an instructional approach that consists of fooling students by hiding educational objectives under game-like techniques. For example, Malone and Lepper (1987) argue that competition makes learning feel like play and stimulates engagement and persistence in the learning activity. In the same vein, Hense et al. (2014) assume that from a self-determination perspective (Ryan & Deci, 2000), collaboration in game-based learning fosters motivation by giving the learner the feeling of being part of a community. In general, competition and collaboration are considered to have a positive impact on students' motivation and, as a result, on students' performance (Burguillo, 2010). Rewards and challenges are recognized to have a positive impact on students' motivation and the power to stimulate engagement. However, Bruckman (1999) underlines the risk that instructional game-based techniques that consist of just covering games over the learning content in order to make the learning content more palatable has become synonymous with the *chocolate-covered broccoli* approach of teaching.

For other authors following the romantic tradition of teaching developed by Froebel (1782–1852), a German pedagogue, play has an intrinsic educational value. Indeed, some studies recognize the importance of game elements that enable the development of a common workspace awareness by making visible to the players the presence and activities of their peers for online games (Leemkuil, de Jong, de Hoog, & Christoph, 2003). Peers' influence, combined with the academic ability for learning, is also recognized as important by researchers. In a study using a collaborative game-based learning environment, Sung and Hwang (2012) found a correlation between the time dedicated to discuss with peers and organize the knowledge and the learning performance. Players' interactions that result from competition or collaboration may have a positive influence (Leemkuil et al., 2003; Sung & Hwang, 2012; ter Vrugte et al., 2015); however, the quality of these interactions is worth considering (Van der Meij, Albers, & Leemkuil, 2011). The influence from peers as powerful factor in acquiring knowledge is based on the fact that collaboration can help students to extend and make explicit their knowledge during game-based learning (ter Vrugte et al., 2015). As a result, the influence of collaboration on learning depends on the epistemic quality of dialogues (Van der Meij et al., 2011). Cheng, Wu, Liao, and Chan (2009) also recognize the positive effect of game competition. Indeed, competition based on well-structured activity can provide a clearly defined goal for students. Thus, there are two different approaches for explaining the influence of competition and collaboration on the learning process. One explanation is more focused on the motivational power of game-based learning through the effect of game mechanics. The second considers that competition and collaboration have a positive influence because they enable the learner to interact with his peers (epistemic interactions) and to develop his performance awareness.

Until now the separated effects of competition and collaboration on learning and motivation were discussed. Different models for game-based learning try to benefit from the positive effect of a combination of collaboration and competition. They describe games where teams of students are competing against each other. In the Johns Hopkins models of cooperative learning, called Student Teams-Achievement Divisions (STAD) and Teams-Games-Tournaments (TGT), students work in 4–5 member teams in which teams receive recognition or other awards based on the learning of all team members (DeVries & Slavin, 1978; Slavin, 2008). These models are considered to be cooperative learning techniques that combine group rewards with individual accountability and thus, collaboration with competition.

However, negative effects of collaboration and competition are also identified. Games that include a time–pressure factor (competition against the system) or competition with adversaries may create an affective state of anxiety and pressure that may inhibit learning (Van Eck & Dempsey, 2002). In addition, due to competition, students are exposed to many social comparison messages which may negatively influence their self-conception, emotions, and actions (Gilbert, Giesler, & Morris, 1995). Regarding collaboration, Ke (2008) also reports an observation where an expert student became a leader rather than a peer tutor in the team, solved questions single-handedly without communicating, and thus interrupted his teammates' cognitive processes. This may explain why collaboration improves positive attitudes but not necessarily promotes performance (Ke & Grabowski, 2007). ter Vrugte et al. (2015) suggest a positive effect of collaboration on learning when competition, and the dominance of one student over another, is not present for below-average students.

From this host of research stems the possibility to use competition as well as collaboration as instructional techniques for the design of educational games. However, competition and collaboration might not be considered to be two different options for game design. Every game is competitive, and collaboration is a core element for game-based learning. In general, studies do focus on direct learning effects and motivation, but little is known about how the collaboration and competition influence the learning process and the players' strategies. In the following, we describe an empirical study dedicated to examine how competition and collaboration were implemented for game design and how they influence students' strategies and the learning process.

9.3 An Empirical Study

In the following, we discuss the impact of competition and collaboration on the learning process through an empirical study into the use of Tamagocours, a game dedicated to teach preservice teachers, the legal rules that apply for the use of digital educational resources (Sanchez et al., 2015).

9.3.1 Tamagocours

Tamagocours (Fig. 9.2) is an online, multiplayer and collaborative Tamagotchi. *Tamagocours* is dedicated to teach preservice teachers the rules (i.e., copyright) that comply with the policies for the use of digital resources in an educational context. At

Fig. 9.2 Tamagocours

the Ecole Normale Supérieure of Lyon, the game is implemented as an online and asynchronous teaching program for approximately 200 students each year (Sanchez et al., 2015). A character (*Tamagocours*) needs to be fed with digital educational resources by a team of 2–4 players. Each team is created automatically and randomly. Each player is represented by an avatar so that the game can be played anonymously. The objective is to make the character healthy by feeding it with digital educational resources and each player can see his teammates choosing the resources from a large database, the format with which these resources are used (collective projection, photocopy, post on the intranet, on a website, etc.) and the consequences of using these for feeding the character (from happiness to food poisoning). Each teammate is allowed to send instant messages to his partners and to discuss the relevance of the chosen resources. The copyright laws that should be respected when it comes to use digital resources for educational purposes. Feeding the Tamagocours provokes feedback that depends on the legal characteristics (creative commons, copyrighted, etc.) and how the resource is used. If these characteristics comply with the copyright policies, the character stays healthy (green color) and the player/learner earns points. Otherwise, the *Tamagocours* gets sick (red color) and dies if fed with too many inappropriate resources. Each of five levels can be replayed indefinitely until the level is completed.

The data collected for this study come from two different methodological approaches. The 193 preservice teachers involved in the study were invited to participate in to an online forum dedicated to comment and discuss the educational properties of the game. The students posted 109 messages. These messages enable to collect the opinions of the students about the relevance of a game-based learning approach in this specific case, and also to examine to what extent the students accept to cooperate and to play the game. The study is also based on recording and analyzing the digital traces produced by the students when they play the game. These digital traces entail the different interactions with the game (clicks on the interface, drag-and-drop operations, etc.) and the instant messages sent by the players to their teammates. This information is automatically recorded and the data collected are uploaded with *Undertracks* (Bouhineau et al., 2013), an open web platform dedicated to analyze, share, and visualize digital traces produced when interacting with Technology Enhanced Learning (TEL) systems. The raw data are previously coded with semantic information—for example, different categories of messages have been identified—and aggregate data are produced. The data analysis entails a *Principal Component Analysis* (PCA) to convert the data into *principal components*, a set of values of linearly uncorrelated variables. PCA is followed by a *Hierarchical Cluster Analysis* (HCA) that enables identification of different categories of players defined by a combination of principal components. The strategies followed by the different categories of players are therefore characterized by a specific set of variables, that is, the actions performed by the player such as checking the characteristics of a given resource before feeding the Tamagotchi, checking the legal documentation, sending instant messages, etc.

9.3.2 *Research Objectives*

The study aims at identifying how the implementation of competition, conflict, and collaboration plays a role in the learning process. We want to understand how collaborative and competitive gameplay should be taken into account for the game design so that learning occurs. This has been done through the analysis of three dimensions. (1) Cooperation with the game is examined through the factors that have an impact on the acceptance of the students to play the game. We want to assess to what extent the students accept to cooperate and to play the game (Sect. 9.3). We also want to find links between the choices made for game design and game implementation, and the consequences of these choices on the player's acceptance to play. (2) A second dimension relates to the conflictual interactions between the students and the game considered as an *antagonist system*. Based on the analysis of the players' strategies, we discuss how these interactions are linked to the learning process (Sect. 9.4). (3) The last dimension consists of the collaborative interactions between teammates and their influence on the formulation of knowledge. We also examine how these interactions participate to the learning process (Sect. 9.5).

9.4 Entering into the Game and Cooperation

In this section, we consider students' commitment as *cooperation with the game*. Based on the collected data, we assess to what extent the students involved in the study accepted to address the challenge offered by the game. This will be done through the analysis of the students' strategies and a discussion about what we call their *appropriation* of the game. We also discuss the factors that influence cooperation and commitment.

9.4.1 *Tamagotchi-Killers*

The data automatically collected during the game sessions show that for some students and for the very first levels of the game supposed to be easy to win, the *Tamagocours* is frequently killed by the use of inappropriate resources. These students have been called *Tamagotchi-killers* as in this case the "death" of the Tamagochi seems to result not from errors made by students but from a conscious willingness. It is difficult to evaluate the number of students involved in such strategy. Indeed, the digital traces collected do not enable to distinguish the Tamagotchi deaths that result from errors and deaths that result from the decision to kill the character on purpose. However, messages posted on the forum and informal interviews with students confirm that some students (approximately 5 %) played, at least during a portion of time dedicated to the experiment, another game than the game they were supposed to play. This borderline case illustrates what Henriot already emphasized when he stated that: "le jeu n'est pas dans la chose mais dans l'usage qu'on en fait"[2] (Henriot, 1969). A game is only a proposal and actual play depends on the acceptation of the player to cooperate and to enter the *magic circle*. As a result, gaming does not result from a top-down process beginning with the design of a game and ending with its adoption by the player. The *lusory attitude* of the player matters and the cooperation of the player involves complex phenomena such as interpretation, reconfiguration, and construction (Raessens, 2005).

9.4.2 Cooperation and Commitment

Not all students decided to kill the Tamagochi and even those who adopted such a strategy shifted to another one so that the different levels of the game were finally completed. Thus, a large majority of the teams (78 among 81) completed the five levels of the game. Only three of them did not manage to complete the very last levels (fourth or fifth). It means that the students accepted to address the challenge offered by the game and were committed to the objective of feeding properly the Tamagotchi. Commitment refers to the idea of emotional and moral engagement. Commitment implies that a student accepts to play a game; it means that he accepts to face the challenge and to follow the arbitrary rules of the game. As a result, commitment can be seen as form of cooperation of a player both with the game designer (who designed the challenge of the game) and other players (teammates or adversaries). Thus, commitment results from cooperation and game appropriation. Gonçalves (2013) proposes a model to describe different levels for game appropriation. The first two levels (*accept* and *test*) mean basic interactions with the digital game interface without a real wish to address the challenge offered by the game. The data collected show no real strategy and a very short time dedicated to "play" without success. Approximately 6 % of the students of our study fall in this category. The third and fourth levels (*make choice* and *anticipate*) mean that the player recognizes that his choices have specific effects that can be anticipated. The data show that strategies are built (these strategies are discussed below). A failure to complete a level is followed by new trials and sometimes by success. However, these students do not share messages to argue and to support their view with their teammates (e.g., by formulating a specific rule that his teammates should follow). This last feature is specific to the highest level of appropriation called *mastery*. At this level, the students recognize the strategies that they explored and they are able to draw arguments for or against a given strategy. At least 74 % of students have reached one of the levels called *make choice* and *anticipate* and 60 % the *mastery* level. These levels have a great importance for the educational dimension of the game because by reaching these higher levels the player shapes strategies and, by doing so, he

[2] "Play does not lay in the artefact but in the use of the artefact."

develops the knowledge needed to address the challenge. Indeed, the knowledge to be learned (the copyright laws) is used by the learner to play and to feed the Tamagotchi. Therefore, game appropriation can be seen as a process which involves the cooperation of the player and leads to sustainable commitment. From a learning perspective, this cooperation entails the acceptation of a challenge: the player recognizes that the issue of the game depends on his own decisions and actions and accepts to take the responsibility to address this challenge. Another dimension of the cooperation of the player lies in the acceptation to follow the rules of the game.

9.4.3 Factors That Influence Cooperation and Commitment

The analysis of the messages sent by the students for the online discussion about their learning experience with *Tamagocours* enables identifying elements that have an influence on the students' cooperation and commitment. It is possible to split these elements into two main categories. The first category relates to the game itself and the second to the context for the use of the game.

The game itself is of course of great importance and the *technical and ergonomic aspects* of the game are, according to the students, one of the first factors to play a role: "difficulty to be connected, non-compatibility with internet Explorer (I don't like to install different browsers on my personal computer only to play to Tamagogours)," "the chat performs badly," or "it is difficult to access to the chat." Others students formulate critics regarding the *gameplay* that is "repetitive," show difficulties to "perceive the change from a level to another," or found "the character ugly and the music annoying." A lot of critics relate to the *integration of the game and the educational content*: "reading legal rules and charts do not really fit with the atmosphere of a game" or "a summary should be available so that we would be able to understand our mistakes." The *collaborative mode* of the game seems to have been widely appreciated even if some students regret that not every player accepted to discuss with the chat. The social interactions permitted by the use of the chat seem to have a positive impact on students' commitment.

It is worth noticing that there is not a consensus among students and while some of them appreciate the game ("after all, this creature was pleasant, its 'yum yum' made me happy, its 'beuarg' broke my heart and I was so scared when it became red"), others complain about the adaptation of the game to its users: "I felt treated as a child" and "games said serious are adapted to primary or secondary students but not to graduate students."

The context for the use of the game is another important aspect. The *time of the year chosen for playing* the game and its *compulsory nature* have been judged inadequate by many students: "this year dedicated to prepare a competitive exam is not the best period especially that it is compulsory." The *nature of game-based learning* as a pedagogy provoked also some critics from students. Some of them do not agree that experiential learning is a relevant approach and consider that "exercises" should be preceded by the teaching of the knowledge: "lacking any previous teaching about copyright it was impossible to succeed without the help of chance." Some of them also express the regret that the game session was not concluded with a *debriefing session*: "Feedback from a teacher seems to me crucial to become aware about what was learned during the game." Some students also conclude: "It was a pitfall to force us to read legal documents."

The factors that influence cooperation and commitment are therefore numerous and complex. These factors result from the game and the choices made for the game design appear to be crucial, but these factors also result from the decision taken for the game implementation. Thus, the success in obtaining the cooperation and commitment of students do not only lie in the hands of game designers but also in the hands of stakeholders and educators.

9.5 Conflict as a First Layer of Play

In this section, we consider the conflictual dimension of the game and its influence on students' strategies. The game is considered to play the role of an *antagonist system* and competition results from an individual conflict with the game, a first layer of play. We discuss how the strategies developed by the student who addresses this conflict are linked with the learning process.

9.5.1 Different Strategies

Though it is possible by playing *Tamagocours* to earn points and though, every time, each player has the possibility to consult the score of his team, the competition merely takes the form of a conflict with the game rather than a contest involving different teams. This results from the fact that the scores of the different teams are not displayed and there is no possibility for comparison. As a result, players compete with the game rather than with other players. Regarding the strategies that they develop during this competition, the *Principal Component Analysis* gives three axes of variables and the *Hierarchical Cluster Analysis* showed that players are split up into five classes of players (Sanchez et al., 2015).

The class called *Efficient players* is the most important regarding the number of students (36 %). Efficient players are students who manage to succeed with few errors and few actions on the interface of the game. They also check the available information for the resources that they use before feeding the Tamagotchi and also consult the legal documentation. A lot of students (26.3 %) have been named *Force-Feeders*. Their main feature is that they frequently do not consult the characteristics of the resources that they feed to the Tamagotchi. They also make a lot of actions on the interface of the game and many mistakes. However, their strategy is not necessary a pure random strategy because some of them choose a format for the use of the resources that is the least constraining as possible. *Prudent students* (20.2 %) almost always check the available information about the resources that they use for feeding the Tamagotchi. They also relatively frequently consult the legal documentation, send few messages, and make few errors. *Talkative students* (12.3 %) are very active both for interacting with the game (a lot of actions) and also in sending messages to discuss about the legislation or about the game itself. *Experts* are rare (2.6 %). They frequently consult the legal documentation and send messages to their teammates either to discuss the legal rules or to give advice about the elements that should be taken into account for selecting the resources. Some students also refused to play. They performed less than 6 actions.

9.5.2 A First Layer of Play

How does conflict arise? How does this conflict influence the learning process? The analysis of strategies followed by students shows that there is a first layer of play which concerns the player as an individual (Sanchez et al., 2015). This first layer of play consists of a situation where individuals take decisions and shape strategies for interacting with the game. A conflict arises from this situation. The game becomes a factor of difficulties and disequilibrium (Balacheff, Cooper, & Sutherland, 1997) and the player's actions depend on his conception (even misconception) of the rules that should be applied for feeding the Tamagotchi. Each action, each decision taken leads instantly to a feedback automatically provided by the game. Therefore, the game plays the role of an *antagonist system*. A novice player makes a lot of errors and the game provides negative feedbacks. The errors made by a novice player lead to a decline, and even to the death, of the Tamagotchi. As a result, the game takes the form of a conflict where the objectives of the player (making the Tamagotchi healthy) are antagonized by the resistance of this system. Then, different strategies take place: being prudent and, somehow, efficient or expert. Or, at the contrary, force feeding the Tamagotchi without a deep commitment with the knowledge that is useful to reach the objective. The game also plays the role of a formative *assessment system* (Gee & Shaffer, 2010) where the player can assess his way of thinking and behaving. As a result, the player is autonomous and his autonomy results both from the freedom to perform according to his own understanding of the objective rather than the expectations of the educator, and from feedback provided by the game.

The game is based on an *intrinsic metaphor* (Fabricatore, 2000). It means that the knowledge that is supposed to be learned is needed to face the conflict. A player needs to master the legal rules for copyright to succeed. For example knowing that 70 years after the death of the author a work is not copyrighted anymore is a key knowledge for selecting resources and feeding the Tamagotchi. Therefore, there is a close connection between the gameplay, based on a feeding metaphor, and the educational dimension of the game. Learning to play and learning the educational content are linked together by the arising conflict and learning is expected to emerge from the experience of the player through an adaptive process. This process results from interactions between the player and the game. For this study, a large majority of teams (78 out of 83) managed to achieve the five levels of the game, and thus, demonstrated that they developed a certain level of expertise regarding the legal rules for the use of digital educational resources.

One of the lessons learned from this empirical study is that designing a game for educational purpose consists of designing an *antagonist system* faced by a player within a conflict. From the player perspective, the knowledge expected to be learned becomes a tool that is essential to address this conflict. As a result, the learning process and the autonomy of the player result from the interactions that take place within this conflict.

9.6 Collaboration and Knowledge Formulation

In this section, we consider collaboration through the analysis of students' online discussions during the game session. These discussions consist of a second layer of play. We examine how this collaborative play influences the learning process through the formulation and the validation of the knowledge that is expected to be learned.

9.6.1 *Validation and Formulation*

The analysis of the strategies shaped by the players also shows that there is a second layer of play that concerns, this time, not the individuals but the different players of the same team. These players are involved in online discussions about the subject to be learned. They have been called *talkative players,* but students from the other categories are also concerned (in particular students called *experts*'). The following table (Table 9.1) illustrates the type of discussion that we recorded. This online discussion about the use of one particular resource called "*à l'école des stéréotypes*"[3] took 4 min 20 s during a play session. In parallel, players performed two types of actions: they checked the characteristics of the resource, that is, the subject of the discussion (eight times) and consulted the legal documentation (player 40 reading about *Creative Commons* license).

The discussion starts with player tagged player 40 (P40). He first checks the characteristic of a resource in the shelf and asks his teammates (P41 and P42) to have a look at this resource. Then, a discussion takes place. First, P42 alerts his teammates about the fact that they have to take into consideration the number of pages of the resource. In other words, he formulates a rule such as: the number of pages matters when it comes to decide if a digital resource is free to use for educational purposes. Following Brousseau's work (Balacheff et al., 1997), such messages dedicated to formulate a rule have been coded as F (*formulation*).

[3] "School of Stereotypes" (a French book written by Christine Morin-Messabel and Muriel Salle).

Table 9.1 Messages excerpt from Team 34

Message Nr	Group Id	User Id	Coding	Messages (translated from French)
1	34	P40	V	Take a look at "à l'école des stéréotypes"
2	34	P42	F	Be careful about the authorized number of pages
3	34	P40	V	Do we keep it or not?
4	34	P42	V	Ok, 25 pages used from "l'école des stéréotypes"
5	34	P42	V	It's a digital edition
6	34	P41	V	So we put it in the fridge, don't we?
7	34	P40	V	Yes
8	34	P42	V	Wait, is there no rule?
9	34	P42	V	It's protected by copyright
10	34	P40	V	A rule??
11	34	P42	V	Not an exception for educational contexts!!
12	34	P41	F	Oh then no if it's digital, we don't have the right.
13	34	P40	F	Yes it's protected by copyright
14	34	P41	F	The document clearly says that we are not authorized to use digital editions
15	34	P42	V	Yeah
16	34	P42	V	But it's a book originally
17	34	P40	V	It's a paper edition right?
18	34	P42	V	It's a scan
19	34	P42	F	So it's authorized
20	34	P42	F	Number of pages limited to 5!
21	34	P42	OM	Sorry
22	34	P40	V	ok so NO

These messages also express the knowledge that is expected to be learned. Other messages dedicated to discuss the validity of these rules (i.e., messages dedicated to ask a question, to make an observation, or to propose hypothesis) have been coded V (*validation*). The messages to express opinions about the game itself or about any other subject are coded OM (*other message*).

The discussion continues and new information is brought mainly by P42: the resource has 25 pages; this digital edition is still protected by copyright… P40 seems to take the role of leader. The data recorded before this episode show that he frequently asks his teammates to have a look at a specific resource and he organizes the selection of the resources ("*let's resume our items one by one*" and "*we must find another one now*"). He also frequently consults the legal documentation (20 times) and usually, he personally takes the decisions to feed the *Tamagocours* or to ask his teammates to do so.

P41 is also involved in formulating ideas about the legal rules that should be respected ("*Yes it's protected by copyright*") but the final argument is provided by P42 ("*Number of pages limited to 5!*") and P40 takes the final decision to not use the resource ("*Ok so NO*") implicitly agreed by the other players.

9.6.2 A Second Layer of Play

The above excerpt shows that the teammates collaborate in order to reach a consensus on rejecting a resource not compliant with the legal rules. This discussion shows that the players share a common objective and that they collaborate. This collaboration involves *epistemic interactions* that consist of two types of interactions: the *formulation* of knowledge that should guide the decisions for acting within the game and observations, assertions, and arguments that are used by the group for the collaborative *validation* of this knowledge. These interactions enable for making explicit the knowledge that is used for individual play. Thus, they might play an important role for the learning process. These recorded interactions are also evidence that the players somehow master the knowledge that they are expected to learn. According to the data collected, 60% of the students who participated in our study managed to write messages dedicated to discuss and formulate the legal rules that comply with the use of digital educational resources. It shows that for these students, the knowledge that they used for feeding the Tamagotchi became explicit and that they were able to communicate this knowledge to their teammates. For the students, this communication enables for a coordinated effort to feed the Tamagotchi and solve the problem together.

We consider this collaborative play as *a second layer of play* important for achieving educational objectives in a game-based learning context. Indeed, the first layer of play enables the development of knowledge, but this knowledge remains mainly implicit. During the collaborative play, the knowledge becomes explicit. The students who managed to formulate legal rules demonstrated that they developed specific knowledge about the criteria that must be applied for selecting resources that comply with copyright. They also developed the capacity to communicate this knowledge to their peers. Therefore, we considered the learning outcomes achieved for these students. The situated knowledge needed to win the game is also shared and somehow validated by an external source, the teammates. We know that the learner needs to learn how to think about a specific domain at a "meta" level (Gee, 2003). Metacognition results from this explicit sharing of knowledge (Balacheff et al., 1997). Formulation and validation that occur during collaboration enable the player to become aware of the knowledge that he uses to play. This is a very important point to consider for game design, especially in a context where few educators are available for a significant number of students which makes it difficult to organize a debriefing session, a crucial step regarding metacognition (Garris, Ahlers, & Driskell, 2002).

9.7 Conclusion: Games as *Coopetitive* Systems

This study underline the importance of choices made for both the design of the game and its implementation. The students' appropriation of a game can be seen as their willingness to cooperate not only with the game designer but also with the educator who is responsible for the implementation of the game. Cooperation takes place when a player accepts to enter the *magic circle* and agree to take on the arbitrary rules of the game and to face the conflict (Huizinga, 1955).

Our findings are also in line with results from previous studies regarding the influence of competition and collaboration on the learning process. A game is a complex combination of conflict/competition and cooperation/collaboration (Salen & Zimmerman, 2004). Depending on how this combination is implemented, it might have a positive influence (Meluso et al., 2012) and game designers should pay specific attention to collaboration and competition as instructional techniques. Indeed, without proper design, negative impacts of employing digital game-based learning approaches could occur (Provenzo, 1992). Villalta et al. (2011) suggest embedding collaboration into the game's functioning mechanics so that the success of the player is conditional to collaborative play. Different game mechanics are linked to collaboration: shared goals, shared puzzles, shared object, and difficulty of the game (El-Nasr et al., 2010). There are also general conditions of conflict and competition in games (Salen & Zimmerman, 2004). The game system should be equitable so that the victory of the winner will have an incontestable value and uncertainty must always remain with regard to the ending (Caillois, 1958). The conflict should involve specific skills (those that are linked to the learning goals) and everything must be removed except the factors involved in the conflict (Salen & Zimmerman, 2004). Player's actions should also be framed by defining rules that are explicit, unambiguous, binding, and shared by all players so that a sense of fairness will be experienced (Salen & Zimmerman, 2004).

Computer games, when appropriately designed, have a potential to positively influence the learning process through the development of *epistemic interactions* when players collaborate. This potential is also under the influence of external classroom goal structures that may foster collaboration or competition (Ke, 2008). It has been recognized that a game that allows the players both to work together and maintains the ability for individual exploration has a positive impact on learning (Inkpen et al., 1995). Different explanations for such a conclusion stem from the literature on game-based learning and also our study. Motivation and commitment depend on the capacity of the game (and the way it is introduced to students) to create social relationships (both collaborative and competitive), conflictual goals, and to foster commitment so that the player accepts to cooperate and to address the challenge offered by the game. The consequence of the player's commitment involves the emergence of a conflict with an *antagonist system* where the player competes and assesses his way of thinking and behaving. By doing so, he individually participates to a *first layer of play* and develops the needed knowledge to succeed. If players get the opportunity to communicate and to play collaboratively, they might also participate to a *second layer of play* where *epistemic interactions* can take place. Thus, validation (arguments and questions to discuss the validity of knowledge) and formulation (of knowledge) play an important role for making explicit the knowledge used to play individually. Therefore, the results of our study are in line with previous studies that already underlined the importance to give the player the opportunity to make his knowledge explicit (ter Vrugte et al., 2015). They are also in line with authors who consider that the positive influence of competition and

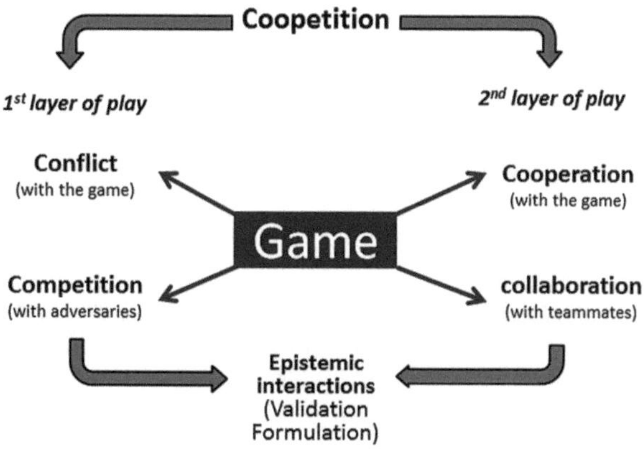

Fig. 9.3 Game as a coopetitive system

collaboration not only results from players' interactions (Leemkuil et al., 2003; Sung & Hwang, 2012; ter Vrugte et al., 2015) but also depends on the quality of these interactions (Van der Meij et al., 2011).

Figure 9.3 is a proposal to describe cooperation/collaboration and conflict/competition as different dimensions of a game and their link with the learning process (*epistemic interactions*). We consider that since collaboration and competition are so intrinsically connected, separating one from the other is impossible. Though a game is more or less competitive or collaborative, there is still always an aspect related to conflict or cooperation in a game. Thus, following the work performed by researchers in economy, we propose to use the word *coopetition*, a neologism formed with the words *cooperation* and *competition*, to describe that a player has to compete and cooperate at the same time. *Coopetition* is a word usually attributed to Ray Noorda and popularized by Nalebuff and Brandenburger (1997). We conclude that games are *coopetitive systems* because they imply dual strategies and these strategies play an important role for learning. Game-based learning consists both of an individual and conflictual play with an antagonist system and a collaborative play which is based on cooperation with the game (to accept to play) and/or with team mates. Learning is thus *coopetitive* as it results from conflictual interactions and argumentative interactions when players collaborate. As a result, instructional techniques used to design games dedicated to educational purposes might take into consideration how competition and collaboration are implemented so that *epistemic interactions*, and learning, result from playing the game.

References

Abt, C. (1970). *Serious games*. New York, NY: The Viking Press.

Alessi, S. M., & Trollip, S. R. (2001). *Multimedia for learning* (3rd ed.). Boston, MA: Allyn & Bacon.

Baker, M. (1999). Argumentation and constructive interaction. In P. Coirier & J. Andriessen (Eds.), *Foundations of argumentative text processing* (Vol. 5, pp. 179–202). Amsterdam, The Netherlands: University of Amsterdam Press.

Balacheff, N., Cooper, M., & Sutherland, R. (Eds.). (1997). *Theory of didactical situations in mathematics: didactique des mathématiques (Didactique des Mathématiques, 1970–1990—Guy Brousseau)*. Dordrecht, The Netherlands: Kluwer Academic.

Bouhineau, D., Lalle, S., Luengo, V., Mandran, N., Ortega, M., & Wajeman, C. (2013). *Share data treatment and analysis processes in Technology Enhanced Learning*. Paper presented at the Workshop Data Analysis and Interpretation for Learning Environments. Alpine Rendez-Vous 2013, Autrans, France.

Bruckman, A. (1999). *Can educational be fun?* Paper presented at the Game Developers Conference '99, San Jose, CA.

Burguillo, J. (2010). Using game theory and competition-based learning to stimulate student motivation and performance. *Computers & Education, 55*(2), 557–566.

Caillois, R. (1958/1967). *Des jeux et des hommes. Le masque et le vertige*. Paris: Gallimard.

Cheng, H. N. H., Wu, W. M. C., Liao, C. C. Y., & Chan, T. W. (2009). Equal opportunity tactic: Redesigning and applying competition games in classrooms. *Computers & Education, 53*(3), 866–876.

Crawford, C. (1982). *The art of computer game design*. Berkeley, CA: Osborne/McGraw-Hill.

DeVries, D. L., & Slavin, R. E. (1978). Teams-games-tournaments (TGT): Review of ten classroom experiments. *Journal of Research and Development in Education, 12*(1), 28–38.

Dillenbourg, P. (1999). *Collaborative learning: Cognitive and computational approaches*. New York, NY: Elsevier Science.

El-Nasr, M., Aghabeigi, B., Milam, D., Erfani, M., Lameman, B., Maygoli, H., & Mah, S. (2010). *Understanding and evaluating cooperative games*. Paper presented at the CHI 2010: Games and Players, Atlanta, GA.

Fabricatore, C. (2000). *Learning and videogames: An unexploited synergy*. Paper presented at the 2000 AECT National Convention, Secaucus, NJ.

Fluegelman, A., & Tembeck, S. (1976). *The new games book*. New York: Doubleday.

Garris, R., Ahlers, R., & Driskell, J. E. (2002). Games, motivation, and learning: A research and practice model. *Simulation & Gaming, 33*(4), 441–467.

Gee, J. (2003). *What video games have to teach us about learning and literacy*. New York: Palgrave Macmillan.

Gee, J., & Shaffer, D. (2010). Looking where the light is bad; video games and the future of assessment. *Edge, 6*(1), 3–19.

Gilbert, D. T., Giesler, R. B., & Morris, K. A. (1995). When comparisons arise. *Journal of Personality and Social Psychology, 69*, 227–236.

Gonçalves, C. (2013). *Appropriation & authenticity—A didactical study on students' learning experience while playing a serious game in epidemiology*. Doctoral dissertation, University of Grenoble, Grenoble.

Hense, J., Klevers, M., Sailer, M., Horenburg, T., Mandl, H., & Günthner, W. (2014). Using gamification to enhance staff motivation in logistics. In S. A. Meijer & R. Smeds (Eds.), *Frontiers in gaming simulation* (pp. 206–213). Stockholm: Springer.

Huizinga, J. (1955). *Homo ludens: A study of the play element in culture*. Boston: Beacon.

Inkpen, K., Booth, K., Klawe, M., & Upitis, R. (1995). *Playing together beats playing apart, especially for girls*. Paper presented at the Computer Support for Collaborative Learning '95 (CSCL), Bloomington, Indiana.

Johnson, D., & Johnson, R. (1996). Cooperation and the use of technology. In D. Jonassen (Ed.), *Handbook of research for educational communications and technology* (pp. 785–811). New York: Macmillan Library Reference.

Juul, J. (2003). *The game, the player, the world: Looking for a heart of gameness*. Retrieved from http://www.jesperjuul.net/text/gameplayerworld/

Ke, F. (2008). Computer games application within alternative classroom goal structures: Cognitive, metacognitive, and affective evaluation. *Education Technology Research & Development, 56*, 539–556.

Ke, F., & Grabowski, B. (2007). Gameplaying for maths learning: Cooperative or not? *British Journal of Educational Technology, 38*, 249–259.

Leemkuil, H., de Jong, T., de Hoog, R., & Christoph, N. (2003). KM Quest: A collaborative internet-based simulation game. *Simulation & Gaming, 34*(1), 89–111.

Lipponen, L. (2002). Exploring foundations for computer-supported collaborative learning. In G. Stahl (Ed.), *Proceedings of the Computer-Supported Collaborative Learning 2002 Conference* (pp. 72–81). Hillsdale, NJ: Erlbaum.

Malone, T. W., & Lepper, M. (1987). Intrinsic motivation and instructional effectiveness in computer-based education. In S. Farr (Ed.), *Aptitude learning, and instruction*. Mahwah, NJ: Lawrence Erlbaum Associates.

Meluso, A., Zheng, M., Spires, H., & Lester, J. (2012). Enhancing 5th graders' science content knowledge and self-efficacy through game-based learning. *Computer & Education, 59*, 497–504.

Nalebuff, B., & Brandenburger, A. (1997). *Co-opetition*. New York: Doubleday.

Ohlsson, S. (1995). Learning to do and learning to understand: A lesson and a challenge for cognitive modeling. In P. Reiman & H. Spade (Eds.), *Learning in humans and machines: Towards an interdisciplinary learning science* (pp. 37–62). Oxford, UK: Elsevier Science.

Parlett, D. (1999). *The Oxford history of board games*. New York: Oxford University Press.

Plass, J., O'Keefe, P., Homer, B., Case, J., Hayward, E., Stein, M., et al. (2013). The impact of individual, competitive, and collaborative mathematics game play on learning, performance, and motivation. *Journal of Educational Psychology, 105*(4), 1050–1066. doi:10.1037/a0032688.

Provenzo, E. (1992). What do video games teach? *Education Digest, 58*, 56–58.

Raessens, J. (2005). Computer games as participatory media culture. In J. Raessens & J. Goldstein (Eds.), *Handbook of computer game studies* (pp. 373–388). Cambridge, MA: MIT Press.

Roschelle, J., & Teasley, S. (1995). The construction of shared knowledge in collaborative problem solving. In C. O'Malley (Ed.), *Computer supported collaborative learning* (pp. 69–97). Heidelberg: Springer.

Ryan, R. M., & Deci, E. L. (2000). Self-determination theory and the facilitation of intrinsic motivation, social development, and well-being. *American Psychologist, 55*, 68–78.

Salen, K., & Zimmerman, E. (2004). *Rules of play, game design fundamentals*. Cambridge, MA: MIT Press.

Sanchez, E., Emin Martinez, V., & Mandran, N. (2015). Jeu-game, jeu-play vers une modélisation du jeu. Une étude empirique à partir des traces numériques d'interaction du jeu Tamagocours. *STICEF, 22*(1), 9–44.

Sanchez, E., Jouneau-Sion, C., Delorme, L., Young, S., Lison, C., & Kramar, N. (2012). *Fostering epistemic interactions with a digital game. A case study about sustainable development for secondary education*. Paper presented at the IOSTE XV International Symposium, La Medina—Yasmine Hammamet, Tunisia.

Schlieder, C., Kiefer, P., & Matyas, S. (2006). Geogames: Designing location-based games from classic board games. *IEEE Intelligent Systems, 21*(5), 40–46.

Slavin, R. E. (2008). Cooperative learning, success for all, and evidence-based reform in education. *Éducation et didactique, 2*(2), 151–159.

Stjerning, J. (2010). *Progress war*. Retrieved February 24, 2016, from http://progresswars.com

Suits, B. (1990). *Grasshopper: Games, life and Utopia*. Boston: David R. Godine.

Sung, H., & Hwang, G. (2012). A collaborative game-based learning approach to improving students' learning performance in science courses. *Computers & Education, 63*, 43–51.

ter Vrugte, J., de Jong, T., Vandercruysse, S., Wouters, P., van Oostendorp, H., & Elen, J. (2015). How competition and heterogeneous collaboration interact in prevocational game-based mathematics education. *Computers & Education*. Retrieved from http://www.sciencedirect.com/science/article/pii/S0360131515300300

Van der Meij, H., Albers, E., & Leemkuil, H. (2011). Learning from games: Does collaboration help? *British Journal of Educational Technology, 42*, 655–664.

Van Eck, R., & Dempsey, J. (2002). The effect of competition and contextualized advisement on the transfer of mathematics skills a computer-based instructional simulation game. *Educational Technology Research and Development, 50*(3), 23–41.

Villalta, M., Gajardo, I., Nussbaum, M., Andreu, J. J., Echeverría, A., & Plass, J. (2011). Design guidelines for classroom multiplayer presential games (CMPG). *Computers & Education, 57*(3), 2039–2053.

Wenger, E. (1998). *Communities of practice. Learning, meaning and identity*. Cambridge, UK: Cambridge University Press.

Chapter 10
Modeling and Worked Examples in Game-Based Learning

Pieter Wouters

Abstract This chapter discusses the role of modeling and worked examples in game-based learning. These instruction techniques can support students to focus on relevant information and to engage in a deeper level of processing (organizing new knowledge and integrate this with prior knowledge) without endangering the motivational appeal of computer games. In addition, it can help to connect informal knowledge representations acquired in the game with the formal domain knowledge representations. Four instructional aspects with respect to modeling and worked examples are discerned: timing (when is the model presented), level of completeness (models can be complete or incomplete), duration (modeling can be faded out or not), and modality (is the model presented in text, pictorial, or both). The results of the review indicate that the use of modeling and worked examples improves learning (d = .61), in particular when it is used to support domain-specific skills, but that little is known about the moderating effect of the four instructional aspects. In addition, it is not clear how modeling and worked examples influence motivation in serious games (d = .01). With respect to both issues more research is required.

Keywords Modeling • Worked examples • Learning • Motivation • Review

10.1 Introduction

The last decade shows an increasing number of empirical studies regarding the effectiveness of computer games in learning, training, and instruction (often referred to as serious games or game-based learning). In addition, reviews and meta-analyses have shown that certain characteristics potentially adhere to game-based learning (GBL) that may hamper learning. One characteristic is that GBL environments can be complex learning environments in which it is not obvious that players automatically engage in processes that yield genuine learning (see Wouters, Van Nimwegen, Van Oostendorp, & Van Der Spek, 2013; Wouters & Van Oostendorp, 2013; Wouters

P. Wouters (✉)
Utrecht University, Utrecht 3512 JE, The Netherlands
e-mail: p.j.m.wouters@uu.nl

P. Wouters, H. van Oostendorp (eds.), *Instructional Techniques to Facilitate Learning and Motivation of Serious Games*, Advances in Game-Based Learning, DOI 10.1007/978-3-319-39298-1_10

& van Oostendorp, this volume). For example, depending on the expertise of the player and the level of sophistication of the game design, players can become overwhelmed by the plentitude of information, the multimodal presentation of information (sometimes simultaneously on different locations of the screen), the choices they potentially can make, the dynamics of the game, and the complexity of the task that has to be performed. This implies that instructional guidance to support players to adequately *select* relevant information and ignore irrelevant information is important, certainly given working memory constraints. The second characteristic pertains to the kind of intuitive learning that is often associated with GBL: players know how to apply knowledge, but they cannot explicate it (Leemkuil & de Jong, 2011; ter Vrugte & de Jong, this volume; Wouters & van Oostendorp, this volume). Yet, it is important that students articulate and explain their knowledge, because it urges them to *organize* new information and *integrate* it with their prior knowledge. These processes will facilitate a knowledge base with higher accessibility, better retention, and higher transfer of learning (Wouters, Paas, & van Merriënboer, 2008). Closely related to this is the observation that players typically acquire informal game knowledge representations that are not adequately connected with the more formal domain knowledge representations (Barzilai & Blau, 2014).

In the remainder of the chapter, I first describe the characteristics of modeling and worked examples. Then I focus briefly on some considerations when using modeling and worked examples in a game-based learning environment. Subsequently, studies investigating modeling and worked examples in GBL are reviewed from a learning and motivation perspective. Finally, some conclusions are drawn and suggestions for further research are presented.

10.2 Modeling and Worked Examples

Both the modeling and the vicarious learning literature emphasize that learning can be enhanced by observing experts or advanced novices display their performance of physical and/or cognitive skills (Bandura, 1976; Collins, Brown, & Newman, 1989; Cox, McKendree, Tobin, Lee, & Mayes, 1999; Van Merriënboer, 1997). When observing an expert perform a complex task in which he/she integrates knowledge and skills, the learner can construct an adequate cognitive representation (Wouters et al., 2008). According to Collins et al. (1989), performance can be divided into the performance of physical skills and processes (e.g., learning to play tennis) and the performance of cognitive skills and processes (e.g., problem solving). The modeling of cognitive skills and processes requires the explication of considerations, thoughts, and reasons that underlie the performance of actions or choices. Whereas physical skills and processes can be observed directly by the learner, considerations in the cognitive domain (e.g., problem solving) are not observable by themselves, especially when abstract processes or concepts are involved. As a consequence these processes need to be externalized, for example, by giving narrated explanations, showing a visual scaffold, etc. (Wouters, Paas, & van Merriënboer, 2008;

Wouters, Tabbers, & Paas, 2007). Worked examples are used in the early stage of skill acquisition and consist of a problem formulation, the solution steps, sometimes important characteristics, and the problem solution (Renkl, 2002). The major difference between modeling and worked examples is that the former not only shows *what* is happening, but also explains *why* this is happening (Collins, 1991; Van Gog, Paas, & van Merriënboer, 2004). In practice, they are closely related and some scholars define worked examples as a type of modeling (Van Gog & Rummel, 2010). Therefore we consider both terms exchangeable.

According to cognitive theories, cognitive resources need to be used efficiently (Mayer, 2008, 2011; Paas, Renkl, & Sweller, 2003; Van Oostendorp, Beijersbergen, & Solaimani, 2008). Human cognitive architecture involves at least two structures (1) working memory with only a limited capacity to process information that is often not sufficient for learning material when it is complex, multimodal, and/or dynamical; (2) long-term memory with a virtually unlimited capacity which can serve as added processing capacity by means of schemas, that is, cognitive structures that can be processed in working memory as a single entity (Kintsch, 1998; Paas et al., 2003). Based on this architecture, cognitive theorists have defined three essential learning processes that are required for learning. Given the limited capacity of working memory, it is crucial that only relevant information is passed for further processing and that nonrelevant information is neglected. The first process, therefore, involves the *selection* of relevant information. The two other processes comprise the *organization* of the incoming information and its *integration* with existing structures in long-term memory (for a more detailed overview, see Mayer, 2011 and Wouters & van Oostendorp, this volume).

Modeling and worked examples are effective because they can support students in all three processes. They can provide initial guidance how to solve a problem (which steps, what characteristics are important, etc.) and thus enable the learner to construct an initial mental model, which can guide learning during task performance. In this respect, the initial mental model will support students to discern relevant from nonrelevant information. Furthermore, modeling can also facilitate the organization and integration of new knowledge with prior knowledge, for example, by gradually removing certain steps from the model. In this way, students are stimulated to generate self-explanations that foster the verbalization of the new knowledge and construct more general applicable knowledge structures (see also the chapter on self-explanations by ter Vrugte & de Jong, this volume). The literature discerns four instructional aspects that can be used to design modeling and worked examples:

10.2.1 Timing

In essence models can be presented before the tasks, after the tasks or just-in-time (Van Merriënboer, Kirschner, & Kester, 2003). When presented before task performance, students start with one or more model(s)/worked examples and then engage in a series of tasks. The purpose of studying the model before task performance is the construction of a mental model that can guide the student during the task. In the just-in-time variant, a model is presented when they are actually required for task performance. The time between the presentation of model and task performance is minimized which enables an optimal use of cognitive resources. Studying the model after task performance can be regarded as a debriefing activity during which game (knowledge) representations can be mapped to the domain representations.

10.2.2 Level of Completeness

Models can be either complete vs. incomplete (Renkl & Atkinson, 2003). Complete models give all steps and the solution of the problem. It supports the student to form an initial mental model that can guide the selection of relevant information for effective task performance. In incomplete models, on the other hand, certain solution steps have been omitted and thus have to be generated by the student. As mentioned earlier, incomplete models are supposed to trigger self-explanations or reflection and generate more robust mental models.

10.2.3 Duration

Duration of the modeling means that modeling can be either faded or not (Renkl & Atkinson, 2003). Fading implies that the nature of the model changes in time, for example, by gradually removing solution steps from the model. Fading is supposed to use cognitive resources more efficiently: students start with a full model that will help them to focus on relevant information, and they will do more themselves as their mental model becomes more developed.

10.2.4 Modality

Instructional information can be presented visually (written text and pictorial aids), verbally (spoken text), or by a combination of both. The choice of one of these formats depends on several factors such as the complexity of instruction (if it is already highly visual, narrated explanations may be more appropriate, see also Kapralos et al., this volume) or the cognitive skills that are involved (some skills such as visuo-spatial skills may be better supported with visual information, see Mayer, Mautone, & Prothero, 2002).

10.3 Modeling in Game-Based Learning Environments

As mentioned before, students typically acquire tacit knowledge during game play that needs to be made explicit in order to create more general knowledge structures that can also be used in situations different from those in the game. Since students are not inclined to do so automatically in games, they need to be stimulated to engage in this level of cognitive processing. Although game-based learning research spans a period of decades, only little value-added research has been conducted in which a game with a model is compared with a game without this model (Mayer, 2011; Wouters & Van Oostendorp, 2013). As we see later, the models in GBL are often external tools—although sometimes embedded in the game environment—that are not integrated in the game play. One of the reasons to use GBL is that the motivational appeal of computer games will also apply to the learning task in the GBL and that this will yield an increase in effort, commitment, and time-on-task. However, when models are not part of the game play, the question can be raised whether these models do not interfere with the game play and thus detract from the motivational appeal of GBL. Therefore, we not only review the effects of modeling and worked examples in terms of learning, but also—when available—their effect on motivation.

The literature search involved the same procedure and criteria as used in the meta-analysis described in Chap. 1. Besides obvious terms like modeling and worked example (and variants such as model or worked-out solution) also terms were included that could refer to modeling/worked example such as 'aids,' 'scaffolds,' and 'concept maps.' For example, in the study of Hwang, Yang, and Wang (2013) concept maps were used. Creating concept maps is neither modeling nor a worked example, but in this study the concept map was embedded in the game which guided the student to collect and organize data. In addition, two different types of learning outcomes were taken into consideration: domain-specific skills (e.g., a knowledge or transfer test) and in-game performance. This division enables some interesting comparisons. Table 10.1 shows the studies that met the criteria with the calculated effect sizes for learning and motivation.

10.3.1 Modeling and Learning

Overall, we found that modeling/worked examples indeed enhance learning ($d = .61$). Furthermore, this improvement pertains to both types learning outcomes that were considered: domain-specific skills ($d = .81$) and in-game performance ($d = .46$). The first study with a model in a game-based learning environment was conducted by Mayer et al. (2002) in the domain of geology. In the game, students learned how to recognize geological structures based on specific surface characteristics. Mayer et al. (2002) used cognitive apprenticeship as a guideline to implement

Table 10.1 Studies involving modeling/worked examples and the effect sizes for learning and motivation

Study	Learning outcome	Comparison	*d* Learning	*d* Motivation
Mayer et al.	Game perf.	Modeling/no modeling (exp. 1)	−.18	na
Mayer et al.	Game perf.	Pict. Aid/no aid (exp. 2)	.56	na
Mayer et al.	Game perf.	Strat. Aid/no aid (exp. 2)	.32	na
Mayer et al.	Game perf.	Strat./pict. Aid/no aid (exp. 2)	.69	na
Mayer et al.	Game perf.	Pict. Aid/ no aids (exp. 3)	.84	na
Mayer et al.	Domain skills	Pict. Aid/no aid (exp. 3)	.99	na
Shen/O'Neil	Domain skills	Worked example/no worked example	1.75	na
Shen/O'Neil	Domain skills	Worked example/no worked example	2.14	na
Lang/O'Neil	Domain skills	Before worked example/no worked example	.06	na
Lang/O'Neil	Domain skills	Just-in-time worked example/ no worked example	.56	na
Sandberg et al.	Domain skills	Modeling/no modeling	.47	na
Hwang et al.	Domain skills	Guiding concept maps/no guiding concept maps	.66	.23
Barzilai/Blau	Domain skills	Scaffold unit before/no scaffold	.41	.04[a]
Barzilai/Blau	Domain skills	Scaffold unit after/no scaffold	−.13	.02[a]
Ter Vrugte et al.	Game perf.	Fading worked example/no worked example	.36	na
Ter Vrugte et al.	Domain skills	Fading worked example/no worked example	.25	na

Note: Column 'Comparison' refers to the names that were used in the studies for modeling/worked examples; d=Cohen's effect size, [a]is mean effect size of two measures (enjoyment and flow). As an indication: Cohen (1988) defined, although hesitantly, effect sizes until d=.2 as small, from d=.2 to d=.5 as medium and from d=.5 as large

models. Especially novices can easily become overwhelmed when they have to perform complex tasks in a serious game. Cognitive apprenticeship can help novices to deal with the amount of cognitive processing that is not related to learning, for example, by partially completing some tasks such as giving verbal descriptions of possible data-gathering strategies that can be used in a complex task. It allows students to spend their cognitive resources to processes that facilitate learning such as organizing, interpreting, and analyzing data. Four versions were compared in three experiments. The base version had no modeling at all. In another version, modeling was defined as an expert who solved the first problem in the game and described her actions (the 'classic' form of modeling). A third version provided students with visual examples of geological structures before they engaged in the game tasks. In the last version, also before the game started, students received information about how they could recognize geological structures. The results show that initially the model group with an expert was not better than the no-model group in game performance. In a second experiment, the no-model group was compared with the group with visual examples, the group with the how information, and a group who received both types of information. It appeared that the visual examples group performed best in the game and that the added value of how information (in the both groups condition) was minimal. The explanation for this is that the game-task requires visuo-spatial thinking which is best supported by the visual examples. The final experiment revealed that the visual examples group was not only superior to a no-model group in game performance but also on a transfer test.

Shen and O'Neil (2006) used worked examples in the domain of problem solving which involved a computer puzzle game that requires the players to find the clues, apply tools and knowledge, and solve the puzzles in order to open safes in a mansion. The authors adopted scaffolding, defined as a collection of tools and resources that can be used by students to assist them with instructional activities (Brush & Saye, 2001), as the guideline for the worked examples. These worked examples comprised a set of sheets with screenshots from the game with text/diagrams describing the components of the safe and a step-by-step procedure to open the safe. The worked example group received these examples before they started with the game. The control group received no worked examples at all. On all learning measures (knowledge maps for content understanding, retention, and transfer) very strong effects in favor of worked examples were found indicating that worked examples indeed improved learning. In a follow-up study, Lang and O'Neil (2008) used the same game, procedure, and measurements to investigate whether the timing of the model was important. The study compared a group without worked examples, a group with worked examples before the game started (comparable with the worked example group in Shen and O'Neil), and a just-in-time worked example group who received the worked example just before it was required for a task in the game. The results indicate that the just-in-time group performed better on content understanding than the group with no worked example or a worked example before. In another publication (Lang & O'Neil, 2011), put forward that no differences were found for the other two measures. Remarkably, compared with the results of Shen and O'Neil (2006), the worked example before group did not perform better than the no-worked example group. The authors argue that presenting the worked example before the game may have caused a temporal split-attention effect which makes it likely that cognitive resources are used for activities such as retrieving unavailable task-related information that do not improve learning. In the just-in-time condition, there is temporal contiguity because the procedural information that is needed to perform the task at that time is also supplied.

In the domain of knowledge management, Sandberg, Wielinga, and Christoph (2012) investigated the effectiveness of prescriptive modeling in a collaborative setting on problem solving (knowledge management). In the game, the students had to manage the knowledge household of a fictitious company. Problems occurred in the form of events, for example, a senior manager leaves the company and starts working for a competitor. This has consequences for the knowledge infrastructure and the students have to deal with it in an effective and efficient way. The focus of the study is the task-layer containing declarative knowledge describing types of problems and the goals (tasks) that constitute a solution to a particular problem type and procedural knowledge describing how a problem can be decomposed and which activities can be used to achieve subgoals. In the context of the game, the task-layer of the prescriptive model described the different steps that need to be taken while solving knowledge management problems. The model was embedded in the game environment and as such retrievable at every moment. In this sense, the model supports a systematic approach to problem solving in the domain of knowledge management. The assumption was that the model would facilitate knowledge acquisition (about knowledge management) and that the model would release cognitive resources because the inclusion of the task model replaced the need to use metacognitive skills. The model group performed better on general and domain-specific procedural knowledge and transfer than the no-model group, but not on declarative knowledge. In addition, it was found that the model group spent nearly twice as much time to finish the game. Although it was assumed that the task model would take over the meta-level, metacognition was still required on the level of collaboration (e.g., discussion on what to do next).

Hwang et al. (2013) used concept maps in the domain of biology. Concept maps were defined as an effective visualized learning tool that helps students memorize and organize their knowledge. It is supposed to help students to engage in high-order thinking and help clarify misconceptions. The use of concept maps as a way of modeling is not obvious, especially when students have to generate concept maps by themselves without any guidance (see Charsky & Ressler, 2011). In this study, however, the concept map is embedded in the game and it guides the student to collect and present data in a well-organized manner. In other words, the aim of these guiding concept maps is to support the student to integrate new learning experiences with their prior knowledge. The results of the study indeed showed that the group with guiding concept maps learned more in terms of transfer. In addition, this group reported a decrease in cognitive load.

The effect of models in the form of scaffolds on solving financial-mathematical word problems was investigated by Barzilai and Blau (2014). They contend that scaffolds may help student to connect the informal knowledge representations that they acquire in the game with the formal knowledge representations of the domain. For this purpose, scaffolds can use two mechanisms: on the one hand scaffolds structure the task and thus reduce its complexity, on the other hand they may problematize the task by causing the students to pay more attention to relevant information that is important for the task (Reiser, 2004). Although problematizing may make the task more difficult in the short term, the authors argue that it will also provide opportunities for deeper processing. In the study, the scaffold was an additional unit in which students learned important concepts (cost, prize, and profit) and their relation, but they also received formal problems that they had to solve. While solving these problems they received informative feedback. Besides investigating the effectiveness of a model, the study also aimed at examining whether the timing of the model would moderate the effectiveness. Presenting it before the task, the model would function as an advanced organizer. Presenting it after the task, the

Fig. 10.1 Complete worked example

model would function as a debrief tool in the sense that it could help students to reflect on what they had learned in the game. Although starting with the scaffold yielded a higher posttest performance compared to the two other groups, the hypothesis that scaffolds would support the explication of intuitive knowledge could not be confirmed because no significant learning gains (posttest–pretest) differences were found. The authors argue that the intervention was rather short (one session) and that a longer intervention may yield significant learning gains for the scaffold groups. The results also suggest that presenting the scaffold before the game (as an advance organizer) is more efficacious than presenting the scaffold after the game (as a review activity).

The last study in this review (ter Vrugte, in preparation) concerns a math game on proportional reasoning. Students received fading worked examples or no worked example at all. In the fading worked examples, group students started with a level in which all information (procedural as well as declarative) was presented that was required to solve the problems. Figure 10.1 shows an example. The required ratio in the worked example is six cola per nine fanta. The problem that has to be solved is how many cola bottles have to be added in case of 30 bottles of fanta when the same proportion between cola and fanta is required? The worked example visualizes this by using the greatest common divisor procedure which means that the required ratio is simplified in such a way that it is easier to calculate with. The simplification is shown in the third column: both numbers can be divided by 3, so the proportion can be simplified to two cola and three fanta. The symbols and texts (by hovering with the mouse) below and above the table give explanatory information showing how the problem is solved. In the subsequent levels, specific solution steps are omitted and requires students to conduct that step themselves. For example, in level 2 the simplification solution step is omitted (see Fig. 10.2).

Fig. 10.2 Partial worked example with the simplification solution step omitted (the *dots* in the third column)

In this study, two interesting observations were made. To start with, the fading worked example group yielded higher learning gains (posttest–pretest) on domain knowledge, but not on transfer. In addition, students were asked to explicate the strategy they used to solve the problems in the posttest and it appeared that the fading worked example group was better capable to verbalize these strategies which may have yielded more abstract and generally applicable knowledge structures (see also ter Vrugte et al., this volume; ter Vrugte et al., in preparation).

10.3.2 Modeling and Motivation

Although perceived motivation was only measured in two studies (see Table 10.1), the results indicate that modeling/worked examples have no negative impact on motivation (the overall corrected effect size is .01). The study with embedded concept maps (Hwang et al., 2013) used seven items to measure motivation, but found no differences between both groups indicating that the concept map may have been well integrated in the game play. In the study by Barzilai and Blau (2014) on scaffolds, two measures were used that give an indication about the motivational appeal: a five-item questionnaire on flow and a three-item questionnaire on enjoyment. Also in this study, there were no differences between the group with scaffolds and the group with no scaffolds indicating that the inclusion of a scaffold did not detract from the motivation. A difference between the Hwang et al. and the Barzilai and Blau study is that scaffold in the latter study was presented either before or after the game and thereby did not interfere with the game play.

10.4 Conclusion and Discussion

Reviews have shown the potential of GBL in education. However, some characteristics come with these games that attenuate this potential positive effect. The fact that games typically facilitate learning-by-doing can lead to more implicit knowledge that is not or only limited applicable in new problems or tasks outside the game context (e.g., in a posttest) (Leemkuil & de Jong, 2011; ter Vrugte & de Jong, this volume; Wouters & van Oostendorp, this volume). In addition, games can be rather complex learning environments which can make it difficult, especially for novices, to discern information that is relevant for the task. Therefore, it is important to implement instructional techniques that support students to select the relevant information and to prompt them to explicate the new learned knowledge so that it can be optimally integrated with their prior knowledge. In this chapter, it is argued that modeling/worked examples, by showing how (and for what reasons) a problem is solved, whether or not, with prompts that trigger students to make explicit their thoughts, can serve these instructional goals. Four instructional aspects were taken from the literature to classify models: timing (when is the model presented), levels of completeness (are some solution steps omitted or not), duration (faded or not), and the modality of the model (visual or textual).

The mean effect size of modeling on learning is $d = .61$ which can be classified as moderate. The studies that have measured motivation give some indication that models do not jeopardize the motivation as long as they are either integrated with the game play or when they are presented outside the context of the game play ($d = .01$), though the number of available comparisons is small. Altogether there are indications that modeling/worked examples can be an effective instructional technique because they support students in GBL to engage in relevant learning processes without interfering with a crucial game characteristic: the motivational appeal. On the other hand, it is difficult to discern which aspects of modeling/worked examples exactly moderate this effectiveness. In Sect. 10.2, four instructional aspects were discerned (timing, completeness, fading, and modality). Although all studies involved modeling, Table 10.2 shows that the studies only focused on one instructional aspect at a time. For example, in the Mayer et al. study, the focus was on the role of modality in modeling, but the role of timing, completeness, and fading were disregarded.

Table 10.2 The use of instructional aspects in the studies involved in this review

	Mayer et al.	Shen/O'Neil	Lang/O'Neil	Sandberg et al.	Hwang et al.	Barzilai/Blau	Ter Vrugte et al.
Timing			X			X	
Completeness					X[a]		X[a]
Fading							X[a]
Modality	X						

[a]Means that these instructional aspects were combined in one condition

Table 10.2 also shows that two studies (Barzilai & Blau, 2014; Lang & O'Neil, 2008) have investigated the effect of timing in model/worked examples. Based on the results in Table 10.1, there is some evidence that the timing consideration of worked example is paramount: just-in-time is better than before and before is better than after, but more research on this consideration as well as on the other consideration is required.

Although the limited number of studies and the extreme effect sizes in some studies should be kept in mind, it seems that the effect of modeling/worked examples is more effective for improving domain-specific skills (e.g., knowledge acquisition and transfer, $d = .81$) than for increasing in-game performance ($d = .46$). Scholars have argued that players of serious games are more likely to learn to play the game (i.e., in-game performance) rather than learn domain-specific skills (Ke, 2009; Leutner, 1993). However, when instructional support was taken into account Wouters and Van Oostendorp (2013) found that serious games with instructional support foster domain-specific skills more than in-game performance. This review reveals the same pattern for modeling/worked examples which may be an indication that this instructional technique indeed supports students to engage in effective organizing/integrating cognitive processes and in this way generate mental models that enable them to solve problems beyond the context of the game.

References

Bandura, A. (1976). *Social learning theory*. Englewood Cliffs, NJ: Prentice Hall.

Barzilai, S., & Blau, I. (2014). Scaffolding game-based learning: Impact on learning achievements, perceived learning, and game experiences. *Computers & Education, 70*, 65–79.

Brush, T., & Saye, J. (2001). The use of embedded scaffolds with hypermedia-supported student-centered learning. *Journal of Educational Multimedia and Hypermedia, 10*, 333–356.

Charsky, D., & Ressler, W. (2011). "Games are made for fun": lessons on the effects of concept maps in the classroom use of computer games. *Computers & Education, 56*(3), 604–615.

Cohen, J. (1988). *Statistical power analysis for the behavioral sciences* (2nd ed.). Hillsdale, NJ: Lawrence Erlbaum Associates.

Collins, A. (1991). Cognitive apprenticeship and instructional technology. In L. Idol & B. F. Jones (Eds.), *Educational values and cognitive instruction: Implications for reform* (pp. 121–138). Hillsdale, NJ: Erlbaum.

Collins, A., Brown, J. S., & Newman, S. E. (1989). Cognitive apprenticeship: Teaching the crafts of reading, writing, and mathematics. In L. B. Resnick (Ed.), *Knowing, learning, and instruction: Essays in honor of Robert Glaser* (pp. 453–494). Hillsdale, NJ: Lawrence Erlbaum.

Cox, R., McKendree, J., Tobin, R., Lee, J., & Mayes, T. (1999). Vicarious learning from dialogue and discourse. *Instructional Science, 27*, 431–458.

Hwang, G.-J., Yang, L.-H., & Wang, S.-Y. (2013). A concept map-embedded educational computer game for improving students' learning performance in natural science courses. *Computers & Education, 69*, 121–130.

Ke, F. (2009). *A qualitative meta-analysis of computer games as learning tools. Handbook of research on effective electronic gaming in education* (Vol. 1, pp. 1–32). Hershey, PA: Information Science Publishing.

Kintsch, W. (1998). *Comprehension: A paradigm for cognition*. Cambridge: Cambridge University Press.

Lang, J., & O'Neil, H. (2008). *The effect of presenting just-in-time worked examples for problem solving in a computer game*. Paper presented at the American Educational Research Association, New York, USA.
Lang, J., & O'Neil, H. (2011). Using computer games to teach adult learners problem solving. In S. Tobias & D. Fletcher (Eds.), *Computer games and instruction* (pp. 435–452). Charlotte, NC: Information Age Publishing.
Leemkuil, H., & de Jong, T. (2011). Instructional support in games. In S. Tobias & D. Fletcher (Eds.), *Computer games and instruction* (pp. 353–369). Charlotte, NC: Information Age Publishing.
Leutner, D. (1993). Guided discovery learning with computer-based simulation games: Effects of adaptive and non-adaptive instructional support. *Learning and Instruction, 3*(2), 113–132.
Mayer, R. E. (2008). Applying the science of learning: Evidence-based principles for the design of multimedia instruction. *American Psychologist, 63*, 760–769.
Mayer, R. E. (2011). Multimedia learning and games. In S. Tobias & J. D. Fletcher (Eds.), *Computer games and instruction* (pp. 281–305). Greenwich, CT: Information Age Publishing.
Mayer, R. E., Mautone, P., & Prothero, W. (2002). Pictorial aids for learning by doing in a multimedia geology simulation game. *Journal of Educational Psychology, 94*(1), 171–185.
Paas, F., Renkl, A., & Sweller, J. (2003). Cognitive load theory and instructional design: Recent developments. *Educational Psychologist, 38*, 1–5.
Reiser, B. J. (2004). Scaffolding complex learning: The mechanisms of structuring and problematizing student work. *The Journal of the Learning Sciences, 13*(3), 273–304.
Renkl, A. (2002). Worked-out examples: Instructional explanations support learning by self-explanations. *Learning and Instruction, 12*(5), 529–556.
Renkl, A., & Atkinson, R. K. (2003). Structuring the transition from example study to problem solving in cognitive skill acquisition: A cognitive load perspective. *Educational Psychologist, 38*(1), 15–22.
Sandberg, J. A. C., Wielinga, B. J., & Christoph, L. H. (2012). The role of prescriptive models in learning. *Computers & Education, 59*, 839–854.
Shen, S.- J., & O'Neil, H. (2006). *The effectiveness of worked examples in a game-based learning environment*. Paper presented at the American Educational Research Association, San Francisco, USA.
ter Vrugte, J., de Jong, T., Vandercruysse, S., Wouters, P., van Oostendorp, H., & Elen, J. (in preparation). Game-based mathematics education: Do fading worked examples facilitate knowledge acquisition?
Van Gog, T., Paas, F., & van Merriënboer, J. J. G. (2004). Process-oriented worked examples: Improving transfer performance through enhanced understanding. *Instructional Science, 32*, 83–98.
Van Gog, T., & Rummel, N. (2010). Example-based learning: Integrating cognitive and social-cognitive research perspectives. *Educational Psychology Review, 22*(2), 155–174.
Van Merriënboer, J. J. G. (1997). *Training complex cognitive skills*. Englewood Cliffs, NJ: Educational Technology.
Van Merriënboer, J. J., Kirschner, P. A., & Kester, L. (2003). Taking the load off a learner's mind: Instructional design for complex learning. *Educational Psychologist, 38*(1), 5–13.
Van Oostendorp, H., Beijersbergen, M. J. & Solaimani, S. (2008). Conditions for learning from animations. In *Proceedings of the* 8th *International Conference of the Learning Sciences* (pp. 438–445). International Society of the Learning Sciences.
Wouters, P., Paas, F., & van Merriënboer, J. J. M. (2008). How to optimize learning from animated models: A review of guidelines based on cognitive load. *Review of Educational Research, 78*, 645–675.
Wouters, P., Tabbers, H. K., & Paas, F. (2007). Interactivity in video-based models. *Educational Psychology Review, 19*(3), 327–342.

Wouters, P., Van Nimwegen, C., Van Oostendorp, H., & Van Der Spek, E. D. (2013). A meta-analysis of the cognitive and motivational effects of serious games. *Journal of Educational Psychology, 105*(2), 249–265.

Wouters, P., & Van Oostendorp, H. (2013). A meta-analytic review of the role of instructional support in game-based learning. *Computers & Education, 60*(1), 412–425.

Chapter 11
Reflections on Serious Games

Arthur C. Graesser

Abstract This chapter comments on the contributions in this edited volume and identifies some challenges for future research on serious games. The contributors used rigorous experimental methods to systematically assess the impact of many components of serious games on learning and motivation. The games are *serious* because there is alignment with relevant instructional content in educational curricula and there is an assessment of associated knowledge, skills, and strategies. The chapters report learning gains for the games compared to comparison conditions, as well as the added value of several game features, such as multimedia, realism, challenge, adaptivity, feedback, interactivity, modeling, collaboration, competition, reflection, fantasy, narrative, and so on. These features are highly correlated in most games so it is difficult to assign credit to particular features when they are implemented in conjunction with many other features. Additional challenges emerge when the games target deep learning of difficult material: (1) game features imposing extraneous cognitive load on working memory, (2) incompatibilities in the timing of feedback to optimize deep learning versus motivation, and (3) control struggles between the game agenda and students' self-regulated learning. It is argued that researchers could be more involved in the building of games under the guidance of scientific principles even though there are difficulties in the design process and in attempts to scale up researcher-designed serious games. The chapter ends with a quandary in assessing psychological constructs in serious games that are adaptive to the learner.

Keywords Deep learning • Game design

A.C. Graesser (✉)
Department of Psychology and Institute for Intelligent Systems, University of Memphis,
202 Psychology Building, Memphis, TN 38152-3230, USA
e-mail: graesser@memphis.edu

P. Wouters, H. van Oostendorp (eds.), *Instructional Techniques to Facilitate Learning and Motivation of Serious Games*, Advances in Game-Based Learning, DOI 10.1007/978-3-319-39298-1_11

11.1 Reflections on Serious Games

The holy grail of serious games is to optimize both learning and motivation. This is a major challenge even though the esteemed *Psychology of Learning and Motivation* edited series has continued to evolve for over 60 years. In truth, the connections between learning and motivation have been sparse when the material to be learned is on difficult academic topics. This is quite apparent when we try to find the learning–motivation connections in serious games. Some of what many researchers and teachers know is intuitively obvious. For example, students will become bored when they perform well on drill and practice games without progressing. There is a negative correlation between learning and enjoyment when the material is difficult and the learners have low knowledge of the subject matter. Game features can reduce learning of academic material when they divert attention from the subject matter. These and other core findings make it apparent that we need to consider the difficulty of the learning material, the proficiencies of the learner, and the features of the game tasks if there is any hope of solving the holy grail of optimizing learning and motivation. The serious games community will hopefully succeed.

My colleagues and I have investigated how serious games can potentially coordinate motivation with deep learning (Graesser, Hu, Nye, & Sottilare, in press; Halpern et al., 2012; McNamara, Jackson, & Graesser, 2010; Millis et al., 2011), as will be addressed throughout this chapter. Deep learning involves systems thinking, reasoning, problem solving, and comprehension of mental models, whereas shallow learning involves memorizing facts, identifying perceptual patterns, and executing simple procedures. Deep learning is more taxing on the limited capacity working memory and executive control processes so it would be expected that deep learning involves more work. That work runs the risk of being incompatible with pleasure and play. However, the complexity would theoretically be pleasurable if it fits within the learner's zone of proximal development, the learner has achieved self-efficacy through mastery of the material, and there is clear, timely feedback on progress (Gee, 2003). Serious games have the potential of turning serious work on important subject matter content into to play (Lepper & Henderlong, 2000). If serious games cannot conquer deep learning, then they offer no solutions for preparing citizens to fill the high paying jobs of the twenty-first Century. Instead, serious games will be relegated to a less lofty niche of entertaining drill and practice on the skills of low paying jobs. Of course, these skills are valuable in their own right and merit engineering solutions to optimize their goals. However, such solutions for shallow learning in games have essentially been solved. Games like *Math Blaster*, *Jeopardy*, and *Angry Birds* can occupy the minds of children and adults for hours.

The big challenge lies in making deep learning engaging and even fun rather than a frustrating struggle (Graesser et al., in press; Rowe, Shores, Mott, & Lester, 2011; Tobias & Fletcher, 2011). The game would need to somehow transcend the negative relationship between deep learning and how much the student likes the learning experience. In most academic environments on difficult topics, thinking hurts. Yet serious games have the potential to shift the learning-liking correlation from negative to positive. Imagine students in a flow experience in the wee hours of the morning trying to solve simulation problems in Science, Technology, Engineering, and Mathematics (STEM), or designing robots, electronic circuits, and cities. They would be engaged for hours, just as they are engaged in frivolous games on topics of no practical consequence.

An ideal serious game to increase deep learning would presumably need to incorporate many of the learning and motivation principles investigated in this edited volume, as well as other edited books on the science of serious games (O'Neil, in press; O'Neil & Perez, 2008; Ritterfeld, Cody, & Vorderer, 2009; Tobias & Fletcher, 2011). The chapters in this volume have used rigorous experimental methods to systematically assess the impact of many components of serious games on learning and motivation for valued academic material: Alignment with relevant instructional content (Vandercruuysse & Elen, Chap. 2; Martinez-Garza, Chap. 3), assessment of cognitive and noncognitive proficiencies (Shute, Ke, & Wang, Chap. 4), fidelity of multimodel interaction (Kapralos, Moussa, Collins, & Dubrowski, Chap. 5), narrative with motivational elements of curiosity and surprise (van Oostendorp & Wouters, Chap. 6), feedback to the learner on their performance (Johnson, Bailey, & Buskirk, Chap. 7), explanations and reflections about the difficult material (Vrugte & de Jong, Chap. 8), collaboration and competition (Sanchez, Chap. 9), and modeling procedures in successful examples (Wouters, Chap. 10). All of these chapters take up the challenge of improving deep learning, as opposed to improving shallow learning or frivolous nonacademic content—the cottage industry of the world of games.

11.2 The Impact of Game Components on Learning and Motivation

Chapter 1 (Wouters and van Oostendorp) provides a comprehensive overview of the science of serious games. It is reassuring to read that serious games improve learning and motivation over and above traditional instruction according to meta-analyses on dozens of studies (see also Clark, Tanner-Smith, & Killingsworth, 2014; Wouters, van Nimwegen, van Oostendorp, & van der Spek, 2013). The chapter also reports effect sizes in improving learning for particular game features, such as multimedia, realism, challenge, modeling, adaptivity, feedback, interactivity, collaboration, reflection, fantasy, narrative, and so on. We learn that some of these features are more important than others in predicting improvements in learning and/or motivation. A major challenge in the latter endeavor is that the features are significantly correlated in most games so it is difficult or impossible to tease apart the added value of specific game features. It is sometimes impossible to vary one feature at a time without seriously compromising the psychological integrity of the game. One challenge for future research is to build quantitative models on the assignment of credit to particular game features in affecting outcome measures of learning and motivation. That being said, the chapters in this volume provide a very informative first-cut approximation of such credit assignments.

Chapter 1 presents a cognitive-affective model as a guiding theoretical framework for the research on serious games (Mayer, 2011; Moreno & Mayer, 2007). This model provides a reasonable foundation for interpreting a large body of research on serious games. The model considers the complexity of the material and tasks, the load on working memory, the dynamic nature of the game interaction, and the knowledge of the learner. A complex game that imposes a high working memory load and confusing series of game events would lead to a frustrating or boring experience for a low-knowledge learner, but possibly an exciting, challenging experience for a high-knowledge learner. According to the model, there is an important distinction between the *selection* of relevant information during learning, the *organization* of the material, and the *integration* of content with prior knowledge. Instruction on the content and game ground rules can facilitate the processes of selection, organization, and integration. Chapter 1 suggests that it is easier to impact selection than organization and integration.

Some game features support entertainment rather than learning, which may make it difficult to coordinate learning and motivation. As argued by Vandercruysse and Elen (Chap. 2), it is preferable to mesh the subject matter content intimately with the game environment (i.e., an *intrinsically integrated game*) than to make subject matter learning an instrumental goal to receiving an extrinsic reward (an *extrinsically integrated game*), or what some call the "chocolate broccoli" approach. Unfortunately, it is extremely difficult to intrinsically integrate the constraints of a complex subject matter with a configuration of game components. For example, imagine a game designer trying to compose a fictive story world (the narrative) that meshes with a scientific mechanism (the academic subject matter). There ideally would be an illuminating harmony between the game narrative and the subject matter content. Unfortunately, the odds of that happening may be comparable to winning an Academy Award or the lottery. Either the narrative does not promote the difficulties of the subject matter, or the narrative is incoherently boring as the designer caters to the constraints of the subject matter. The odds are that the two worlds are in collision unless a very talented design team can find clever ways to connect them. That is perhaps why serious games to promote deep learning of complex systems typically resort to *gamification* with extrinsically integrated game components.

Graesser et al. (in press) identified some other potential liabilities in meshing game components with deep learning of complex subject matters. These are discussed below.

Extraneous nongermane cognitive load. The game components may impose an extraneous cognitive load that is not germane to the mastery of the subject matter. Adams, Mayer, McNamara, Koenig, and Wainess (2012) reported evidence for narrative reducing learning compared to a more conventional instructional method, for example. It is uncertain, however, whether the narrative imposes a short-term penalty (only the first 1–2 h) or a long-term penalty (after ten or more hours).

After the extraneous game components are mastered and automatized, there may be no learning penalties but large advantages from the motivational game components. Jackson and McNamara (2013) reported such a pattern, even when time on task was controlled. They compared a game on self-explanation comprehension training for reading (a motivationally enhanced, *ME-iSTART*) with an intelligent trainer for self-explanation comprehension training (*iSTART*). Both of these systems had collaborative agents (talking heads) that interacted with the students in conversation and guided the self-explanations. The ME-iSTART had game-based competition and performance scores in addition to self-explanations, feedback, and conversational interactions. These features were all addressed in chapters in this volume, namely self-explanations (Vrugte & de Jong, Chap. 8), interactive collaboration and competition (Sanchez, Chap. 9), feedback (Johnson, Bailey, & Buskirk, Chap. 7), and modeling of self-explanations on examples (Wouters, Chap. 10). It was the competition, game performance scores, and choice options that distinguished the ME-iSTART from the iSTART system. Jackson and McNamara (2013) reported that the game components had a short-term penalty but a long-term advantage over time.

The Jackson and McNamara results raise questions about the selection of studies to include in meta-analyses and reviews of the impact of games on learning. The extreme position is to include only those studies in which the students experience the game and comparison conditions for ten or more hours. At hour 10, the experience with the system is sufficiently long that the extraneous game components are understood, if not automatized, and the novelty rush of the game has worn off. A more moderate position is to manipulate or measure the amount of time the student interacts with the game or comparison condition.

Interestingly, available studies have rarely (if ever!) collected perhaps the most important measure of learning that examines the students' behavior when they intrinsically want to use the learning environment. The most important, ideal measure would be a confluence of learning and motivation. One could do this by allowing the students to use the system (game versus comparison condition) for as long as they chose to use the system. Performance, voluntary use, and learning could be tracked over time. Consider the situation in which the game is only half as effective as a traditional comparison condition in performance and learning scores within the first hour, but the student voluntarily uses the game for 10 h in the game condition and 1 h in the comparison condition. We need a metric that computes the time on task for serious content over the temporal journey of using the system voluntarily (perhaps a cumulative score or an integral calculus metric). Without such a metric, we are losing the essence of the value of serious games. It is widely acknowledged that today's students are using games for many hours a day at the expensive of very little time working on homework on academic content. A confluence metric is absolutely essential for future research on serious games. What do students do when they have a choice and are self-regulated? Instead of a metric of learning-per-time-increment during early phases, we need a metric of learning throughout the life cycle of using a learning environment voluntarily.

Feedback timing clashes. Feedback is an important aspect of both games and deep learning. However, the timing and nature of the feedback is different for the two worlds. Games often provide timely, if not quick feedback to the learner about the quality of their contributions in order to keep the student in what Csikszentmihalyi (1990) has called the state of psychological *flow*. Flow is intense engagement to the point where time and sometimes fatigue psychologically disappear. For deep learning, however, there needs to be time for thought and reflection over the depth of the material, a timing pattern that might clash with the speedy tempo of existing games.

The chapters in this volume uniformly comment on the importance of feedback in serious games (particularly Johnson, Bailey, & Buskick, Chap. 7). Feedback is important for both learning and motivation. However, the timing allegedly is very different for deep learning and motivation. Deep learning takes time for reflection and sometimes collaboration (measured in minutes), but motivational feedback is short (measured in seconds). A game presents a challenge, question, or event to the student, followed by the collection of data on how the student responds. Short response times of feedback (whether quantitative or qualitative) optimizes motivation, but longer response times of feedback optimize thinking, reasoning, and reflection. There is an issue of how the game designer resolves these two incompatible constraints.

Control struggles. There is the idealistic vision of the students wanting to be in control and follow their whims in a learning trajectory that is guided by intrinsic motivation and self-regulated learning. Students with higher knowledge and ability benefit from this control whereas novices are lost and their decisions are suspect (Graesser & McNamara, 2010; Hacker, Dunlosky, & Graesser, 2009). Such high ability students are rare, not mainstream, but the future may change in the digital era with Google and widespread use of serious games. We know that self-regulated inquiry is not well developed in most students, but those with high ability might not have the patience for a serious game that is guided by an instructional agenda. This control issue needs to be addressed in assessments of serious games.

11.3 Building Science-Based Serious Games

Imagine an enterprise where researchers apply the science of learning to build serious games to optimize learning and motivation on various subject matters. This approach has been pursued in various research centers, thanks to funding from the Federal agencies and private foundations. It takes considerable resources to build these systems ground up, given that a typical successful commercial game costs $10 million or more.

The chapter by Shute, Ke, and Wang (Chap. 4) describes their experience in building a game (*Physics Playground*) from scratch to apply their 9-step evidence-centered design (ECD) approach. The authors describe how ECD can be set up to assess problem solving stages, creativity, persistence, and other psychological characteristics as students interact with the game, with dozens/hundreds of observations per hour to feed into the assessment. Moreover, the accuracy of the assessment improves over time with feedback from the performance, using modern machine learning methods that dynamically modify the underlying Bayesian networks linking performance to proficiencies. The set of proficiencies and the links from performance to proficiencies are based on scientific principles of learning and motivation. That is a very different approach than examining an existing game that is built from game developers (based on art more than science), identifying underlying principles, and testing different versions of the game that manipulate the principles.

The difficulty of developing a serious game cannot be overstated. There are high expectations of gamers who have experienced the benefits of the multibillion dollar industry of entertainment games. The chapter by Kapralos et al. (Chap. 5) analyzed the extent to which viewers' psychological experience is affected by different dimensions of fidelity to the real world. Some dimensions need not be perfect in order to be engaging and to create a psychological experience of *presence* in the world. Nevertheless, the standards of most viewers are high and the rendering of the perceptual images and actions is expensive. The performance and perceptions of the visual aspects of the game are also influenced by the sound in the multimodal interaction. Classical music was found to help but noise hurt the perceived quality of the experience as well as performance in the serious game. When a modality has low fidelity, it runs the risk of placing an extraneous cognitive load on the student that is not needed to complete the academic task.

Aside from the multimedia fidelity and experiential presence, the designer needs to have events, episodes, actions, and tasks aligned with good pedagogy, the academic curriculum, and subject matter content. Thus, there is a complex mapping between game events, pedagogical principles, and knowledge components in the curriculum (Koedinger, Corbett, & Perfetti, 2012; Shute & Ventura, 2013). Setting up this alignment is a very detailed, if not tedious, activity. The researcher needs to prepare a spreadsheet that specifies how each game task and performance indicator is aligned with particular learning principles and knowledge components in a curriculum. An unmotivated task/indicator would need to be deleted if one were to optimize the efficiency of the game. Moreover, a game is incomplete to the extent that particular knowledge components are not covered and inefficient to the extent to which important learning principles are not applied. This is a long distance from the creative impulses of the game designers of the entertainment industry.

Another part of the process is conducting pilot testing and feasibility studies on groups of learners in order to confirm or discover what aspects of the games are successful. That requires several cycles of iterative testing on small groups of students. The data collection may involve behavioral observation, eye tracking, analysis of computer logs, think aloud protocols, and ratings of student impressions. This design process takes time and follows a systematic methodology, which again is very different than the process of creating games in the commercial world.

My research groups have had mixed success in creating serious games. Each of these efforts was guided by a collection of theoretical frameworks that allegedly explain learning and motivation. In addition to the cognitive-affective model of Mayer and Moreno (Mayer, 2011; Moreno & Mayer, 2007), we were inspired by

Lepper's model of intrinsic motivation (Lepper and Henderlong 2000), some timely books on video games and learning (Gee, 2003; Shaffer, 2007), and our own research on moment-to-moment emotions during deep learning (D'Mello & Graesser, 2012; Graesser & D'Mello, 2012). We never expected a single theory to go the distance in explaining learning and motivation in serious games at that pre-paradigmatic stage of the enterprise. Instead, we considered a collection of theoretical perspectives and associated predictions to guide game design. And indeed, there were often contradictions and trade-offs in the formulation of several predictions. Two cases illustrate some of the difficulties in creating successful serious games.

A serious game learning community at a university. For two years, some colleagues and I organized learning communities to create serious games at my university (Graesser, Chipman, Leeming, & Biedenbach, 2009). Each year, 25–30 students took a game design class plus courses in introductory psychology, English composition, and computer science. All of the students took the same set of classes for a semester. The students were divided into groups of 4–5 with the goal of creating a serious digital or board game in some area of psychology of their choosing. Four graduate students and four professors (one from each course) helped scaffold their education and design activities. They were assigned the popular texts on serious games by Salen and Zimmerman (2004) and Gee (2003) in the game design course, whereas they received the typical curriculum in psychology, English composition, and computer science. In addition to the lectures and course assignments, students had the opportunities to present their ideas, write proposals, and listen to feedback.

The disappointing news is that the games were not very interesting or academically deep. The students resorted to popular game platforms that pitched shallow learning, such as *Jeopardy*, *Trivial Pursuit*, and *Monopoly*. One of the more interesting games was "Psychopoly," which had some creative, psychologically relevant labels for hotels and a "get out of asylum free card," but otherwise was a game that only supported the learning of shallow trivial details. The one positive note is that some of the groups of students continued to meet and expand their game activities after the semester was completed. It is unlikely that students would do that after a course in statistics, research methods, or even introductory psychology.

A commercialized game on scientific reasoning: *Operation ARA*. Some colleagues and I created a serious game to train high school and college students the fundamentals of scientific reasoning (Cai et al., 2011; Millis et al., 2011), thanks to funding from the Institute for Education Sciences. The original system was called *Operation ARIES*! ARIES is an acronym for *Acquiring Research Investigative and Evaluative Skills*. It is an intelligent tutoring system that teaches critical thinking and scientific reasoning. The system has three modules: interactive text (called *training*), *case study*, and *interrogation*. The ARIES training module has an eBook, multiple-choice questions, and tutorial *trialogs* (two computer agents interacting with the human, see Graesser, Li, & Forsyth, 2014). The eBook, *The Big Book of Science*, covers 21 topics or principles of research methodology, such as hypothesis, operational definitions, sampling, independent and independent variables, control groups, and correlation versus causation. After the student masters the training, they move to a second phase that presents a series of cases of scientific studies and the student critiques each case as to whether there are flaws in the scientific design, evidential claims, and reasoning. They are expected to apply the 21 principles in the critiques of cases. In the final interrogation phase, the student generates questions to troubleshoot whether a case study has a flaw. There is a narrative associated with the game that involves aliens invading earth and disseminating bad science so they can take over the planet. The student is being trained by the Federal Bureau of Science to identify and capture these aliens in order to stop the spread of bad science.

The research team was excited about the 20-h game that applied both principles of game design and principles of learning. We were reassured that the game improved learning in pretest → game → posttest designs and in pretest → game vs. comparison condition → posttest designs. We had the good fortunate of commercializing the game through Pearson Education, where the game was implemented on the Internet under a new title *Operation ARA*. The marketers at Pearson insisted that the game be shortened to 6–8 h so various reductions were made, such as fewer chapters and fewer cases. ARA fortunately produced learning gains, much to our relief (Forsyth, Graesser, Pavlik, Millis, & Samei, 2014; Halpern et al., 2012).

There were several disappointments in the ARA project that are instructive. One disappointment was that Pearson did not allow us to modify the software after data were collected from participants who purchased the system. Our understanding was that iterative changes would be made to the game components, assessment parameters, and learning model as we conducted data mining analyses, following the principles of evidence-centered design (see Chap. 4 by Shute, Ke, and Wang; Shute & Ventura, 2013). That was not granted because Pearson was being reorganized with a major change in leadership. This underscores the importance of clearly specifying the design and revision process for research teams who want to participate in building serious games. Indeed, there were other opposing views between the Pearson and the researchers on the design of ARA. For example, the researchers wanted the students to critique many cases in order to automatize scientific reasoning and adequately learn all of the principles of science; the Pearson folks wanted to shorten the time and have a small number of cases to satisfy the economics of marketing the game.

A second disappointment was the feedback that we received from students who played the game. For example, I incorporated the game as part of the curriculum in a large introductory psychology class with 300 students. Completing the game was 10 % of the grade. Some students enjoyed playing the game, but other students found the game to be difficult and were frustrated when there were occasional glitches in the system (which we wanted Pearson to fix but that was not granted). Perhaps this is not a surprising result, given that we know that enjoyment is often negatively correlated with deep learning, as discussed earlier. Moreover, making the game a requirement is very different from students voluntarily choosing to play the game out of intrinsic interest. Again, perhaps this is not a surprising result, given what is known in marketing research (i.e., there are large differences in customer preferences) and given what we know from psychology (i.e., differences between intrinsic and extrinsic motivation).

A third disappointment was that it was difficult to determine how learning and motivation was influenced by particular game components and particular scientific principles of learning or instruction. The game and pedagogical features were highly correlated as bundles of features, so it was difficult to assign credit to the added value of particular features. Forsyth et al. (2014) applied mixed effects modeling in order to partial out how much learning is predicted by different factors, such as time on task, student generation of information, receiving accurate discriminations, amount of scaffolding, student question generation, reading text, exposure to cases, and so on. Most of these factors had a small incremental impact over and above all of the other factors, but the results were not consistently significant across phases of the game and prior knowledge of the students. When many factors exert their impact on learning, the impact of any one factor tends to be small or modest (i.e., getting lost in the crowd). We conducted a lesion experiment that manipulated the presence versus absence of each of the three phases of the game ($2 \times 2 \times 2 = 8$ conditions). One drawback of that approach is that deleting a phase of the game ended up creating some confusion in the narrative, so any decrease in performance from the lesion could be attributed to confounding variables from the manipulation. Just as correlational studies have the risk of being explained by extraneous third variables, true experiments that manipulate variables have the risk of being explained by unwanted confounding variables from the manipulation (Graesser & Hu, 2011). These well-known trade-offs in correlational versus experimental designs apply to game research just as they apply to other areas of psychology.

This latter concern about credit assignment of game features on learning and motivation would apply to most of the chapters in this volume. Some game features are highly separable (i.e., orthogonal) from other features, whereas others are integral (i.e., correlational, interactive, and mutually constraining) with the psychological impact of other features. One might intuitively expect that visual fidelity (texture, degree of resolution, and 2D versus 3D) would be entirely separable from the auditory modality (classical music, noise, and sounds meaningfully coordinated with events), but the chapter by Kapralos et al. (Chap. 5) suggests that the visual detail is perceived to increase when the audio channel is enhanced. Hence, the visual and audio modalities are correlated in the impact on psychological realism. There is uncertainty in which channel gets credit when the two modalities are correlated.

The interactions among game features and individual differences can become very complicated. The chapters in this volume focused on some theoretically informative interactions among game features, individual differences, and pedagogical principles. For example, Oostendorp and Wouter's chapter (Chap. 6) on narrative informs us that the curiosity manipulation did not have much impact on learning, whereas surprise from unexpected events had a larger impact, primarily for higher ability learners with respect to subject matter knowledge and metacognition. Such aptitude–treatment interactions are very important to document. There is also another level of assessing the generality of such interactions across different emotions and game contexts. An adequate theoretical model needs to predict and explain such higher order interactions.

The authors of the edited volume were explicit in claiming that it was beyond the scope of the current research to provide a full account of the many interactions and game contexts. That is prudent at this early stage of serious game research.

However, it should also be acknowledged that the field needs to have a broader landscape of serious games on different subject matters with different populations and game features before there can be confidence in the conclusions. That can only be accomplished if researchers are part of the design process in building serious games. Testing other people's games is inherently limited because of the high density of feature correlations, aptitude–treatment interactions, selection biases on who plays the game, and limited contextual scope. Experimental manipulations are a start, but still limited by the unexpected confounding variables that accompany all experimental manipulations (Graesser & Hu, 2011).

11.4 A Quandary in Stealth Assessment in Adaptive Serious Games

Many are intrigued with two directions in the future of serious games. The first is *stealth assessment* (Chap. 4; Shute & Ventura, 2013). That is, serious games could assess performance, learning, proficiencies on academic topics, personality characteristics, and a broad profile of other psychological attributes without the student even knowing that they are being assessed. The number of moments of assessments would be large (dozens, hundreds, or even thousands per hour) in systems that track student emotions and physiological responses (Calvo & D'Mello, 2010; D'Mello & Graesser, 2010). Moreover, we live in an era of big data in education with the volume, velocity, variability, and veracity of data on increased trajectories (Dede, 2015). Imagine a world where teachers have students playing serious games, with student profiles being continuously updated, a small number of high-stakes tests, and attempts to optimize student motivation so that attentional or physical attrition does not add serious measurement error. Imagine a world where students could look at their personal profile and improve their learning in a self-regulated manner. Imagine teachers and parents inspecting the profiles and trying to help them improve. But also imagine a worried population that Big Brother has taken over and that there is a serious invasion of privacy. Perhaps only the student should benefit from stealth assessment and be the only one who can inspect these data.

The second direction is developing *adaptive* serious games on deep learning. All game theories advocate generating challenges that are tuned to the students' abilities and giving feedback on their performance. Adaptive generation and feedback requires intelligent algorithms. These adaptive mechanisms can be delivered by combining intelligent tutoring systems with serious games, but the advance creates some challenges from the standpoint of value-added in the assessment of particular game features. As a student becomes more accomplished, the student will be assigned more difficult problems and it becomes more difficult to have high scores. The performance on any item in a game is therefore a product of both the learner ability and the difficulty of the item. This makes it complicated in scoring performance of a student.

The complex interactions among learner abilities, item difficulty, and learning factors presented difficulties in assigning credit to game components and learning factors in ARIES and ARA. Consider the comparison of open-ended verbal responses and the selection of options from a menu of alternatives (e.g., explanations and answers to questions). Such comparisons occurred in ARA/ARIES as well as the games investigated by Vrugte and de Jong (Chap. 8). There is a comparison of performance between the alternatives of open-ended responses and the selection from options in a multiple-choice format or through a hint question to elicit a specific idea; Vrugte and de Jong called these *open-ended* versus *directive* treatments, respectively. The affordances of the open-ended treatment is to promote and assess (a) the student's active generation of information as well as (b) time on task in most contexts because it takes effort to compose responses, but at the expense of (c) having less scaffolding and (d) lower discriminations among fine points that are manifested in the directive options. Consequently, any comparison between the open-ended and directive treatments needs to consider the theoretical components of a, b, c, and d. This illustrates the importance of the complex mapping between the game features and theoretical components.

The picture is complicated further by the adaptive assignment of game interactions that are sensitive to the learner's abilities. In ARIES/ARA, for example, the higher ability students experience more events with open-ended responses, but less scaffolding and fewer options to perceive fine-grained discriminations. The higher ability students also receive more difficult problems, following the principle of the zone of proximal development. These mechanisms follow the normal mechanisms of intelligent tutoring systems (VanLehn, 2006) and serious games (Shute & Ventura, 2013). Thus, there is a selection bias in the delivery of game experiences that is sensitive to the learner and that makes it very difficult to assign credit to the game features versus underlying psychological constructs that address learning and motivation.

Complex mathematical models may be needed to ferret out the impact of particular game and pedagogical features on learning and motivation. The value-added of a particular feature is quite complex because of the complex correlations, interactions, and trade-offs inherent in the large set of factors. The notion of assessing one factor at a time (not present versus present) with the remaining factors being not present is impractical, if not impossible, with little hope of generalization. When a person is starving, an apple can have a large impact on his/her emotion, attention, and actions; when the apple is added to a typical rich diet of a person, the incremental impact of an apple is very small. This analogy applies to the study of serious games as well.

Acknowledgments The serious games developed in the Institute for Intelligent Systems at the University of Memphis were support by the National Science Foundation (ITR 0325428, DRK-12-0918409, and DRK-12-1108845) and the Institute of Education Sciences (R305B070349; R305C120001). The opinions, findings, and conclusions do not reflect the views of the funding agencies, cooperating institutions, or other individuals.

References

Adams, D. M., Mayer, R. E., McNamara, A., Koenig, A., & Wainess, R. (2012). Narrative games for learning: Testing the discovery and narrative hypotheses. *Journal of Educational Psychology, 104*(1), 235–249.

Cai, Z., Graesser, A. C., Forsyth, C., Burkett, C., Millis, K., Wallace, P., et al. (2011). Trialog in ARIES: User input assessment in an intelligent tutoring system. In W. Chen & S. Li (Eds.), *Proceedings of the 3rd IEEE International Conference on Intelligent Computing and Intelligent Systems* (pp. 429–433). Guangzhou: IEEE Press.

Calvo, R. A., & D'Mello, S. K. (2010). Affect detection: An interdisciplinary review of models, methods, and their applications. *IEEE Transactions on Affective Computing, 1*, 18–37.

Clark, D., Tanner-Smith, E., & Killingsworth, S. (2014). *Digital games, design and learning: A systematic review and meta-Analysis*. Menlo Park, CA: SRI International.

Csikszentmihalyi, M. (1990). *Flow: The psychology of optimal experience*. New York: Harper-Row.

D'Mello, S., & Graesser, A. C. (2010). Multimodal semi-automated affect detection from conversational cues, gross body language, and facial features. *User Modeling and User-Adapted Interaction, 20*, 147–187.

D'Mello, S. K., & Graesser, A. C. (2012). Dynamics of affective states during complex learning. *Learning and Instruction, 22*, 145–157.

Dede, C. (2015). *Data-intensive research in education: Current work and next steps*. Computer Research Association. Retrieved from http://cra.org/cra-releases-report-on-data-intensive-research-in-education/

Forsyth, C. M., Graesser, A. C., Pavlik, P., Millis, K., & Samei, B. (2014). Discovering theoretically grounded predictors of shallow vs. deep- level learning. In J. Stamper, Z. Pardos, M. Mavrikis, & B. M. McLaren (Eds.), *Proceedings of the 7th International Conference on Educational Data Mining (EDM 2014)* (pp. 229–232). Honolulu, Hawaii: International Educational Data Mining Society.

Gee, J. P. (2003). *What video games have to teach us about language and literacy*. New York: Macmillan.

Graesser, A. C., Chipman, P., Leeming, F., & Biedenbach, S. (2009). Deep learning and emotion in serious games. In U. Ritterfeld, M. Cody, & P. Vorderer (Eds.), *Serious games: Mechanisms and effects* (pp. 81–100). New York, London: Routledge, Taylor & Francis.

Graesser, A.C., & D'Mello, S. (2012). Emotions during the learning of difficult material. In. B. Ross (Ed.), The psychology of learning and motivation (Vol. 57, pp. 183–225). Amsterdam, The Netherlands: Elsevier.

Graesser, A. C., & Hu, X. (2011). Commentary on causal prescriptive statements. *Educational Psychology Review, 23*, 279–285.

Graesser, A. C., Hu, X., Nye, B., & Sottilare, R. (2016). Intelligent tutoring systems, serious games, and the Generalized Intelligent Framework for Tutoring (GIFT). In H. F. O'Neil, E. L. Baker, & R. S. Perez (Eds.), *Using games and simulation for teaching and assessment* (pp. 58–79). Routledge: Abingdon, Oxon.

Graesser, A. C., Li, H., & Forsyth, C. (2014). Learning by communicating in natural language with conversational agents. *Current Directions in Psychological Science, 23*, 374–380.

Graesser, A. C., & McNamara, D. S. (2010). Self-regulated learning in learning environments with pedagogical agents that interact in natural language. *Educational Psychologist, 45*, 234–244.

Hacker, D. J., Dunlosky, J., & Graesser, A. C. (Eds.). (2009). *Handbook of metacognition in education*. Mahwah, NJ: Erlbaum/Taylor & Francis.

Halpern, D. F., Millis, K., Graesser, A. C., Butler, H., Forsyth, C., & Cai, Z. (2012). Operation ARA: A computerized learning game that teaches critical thinking and scientific reasoning. *Thinking Skills and Creativity, 7*, 93–100.

Jackson, G. T., & McNamara, D. S. (2013). Motivation and performance in a game-based intelligent tutoring system. *Journal of Educational Psychology, 105*, 1036–1049.

Koedinger, K. R., Corbett, A. C., & Perfetti, C. (2012). The Knowledge-Learning-Instruction (KLI) framework: Bridging the science-practice chasm to enhance robust student learning. *Cognitive Science, 36*(5), 757–798.

Lepper, M. R., & Henderlong, J. (2000). Turning "play" into "work" and "work" into "play": 25 years of research on intrinsic versus extrinsic motivation. In C. Sansone & J. M. Harackiewicz (Eds.), *Intrinsic and extrinsic motivation: The search for optimal motivation and performance* (pp. 257–307). San Diego, CA: Academic.

Mayer, R. E. (2011). Multimedia learning and games. In S. Tobias & J. D. Fletcher (Eds.), *Computer games and instruction* (pp. 281–305). Charlotte, NC: Information Age.

McNamara, D. S., Jackson, G. T., & Graesser, A. C. (2010). Intelligent tutoring and games (ITaG). In Y. K. Baek (Ed.), *Gaming for classroom-based learning: Digital role-playing as a motivator of study* (pp. 44–65). Hershey, PA: IGI Global.

Millis, K., Forsyth, C., Butler, H., Wallace, P., Graesser, A., & Halpern, D. (2011). Operation ARIES! A serious game for teaching scientific inquiry. In M. Ma, A. Oikonomou, & J. Lakhmi (Eds.), *Serious games and edutainment applications* (pp. 169–196). London, UK: Springer.

Moreno, R., & Mayer, R. (2007). Interactive multimodal learning environments. *Educational Psychology Review, 19*, 309–326.

O'Neil, H. F., Baker, E. L., & Perez, R. S. (Eds.), *Using games and simulation for teaching and assessment.* Routledge: Abingdon, Oxon.

O'Neil, H. F., & Perez, R. S. (Eds.). (2008). *Computer games and team and individual learning.* Amsterdam, The Netherlands: Elsevier.

Ritterfeld, U., Cody, M., & Vorderer, P. (2009) (Eds.), Serious games: Mechanisms and effects. New York and London: Routledge, Taylor & Francis.

Rowe, J. P., Shores, L. R., Mott, B. W., & Lester, J. C. (2011). Integrating learning, problem solving, and engagement in narrative-centered learning environments. *International Journal of Artificial Intelligence in Education, 21*, 115–133.

Salen, K., & Zimmerman, E. (2004). *Rules of play: Game design fundamentals.* Cambridge: MIT Press.

Shaffer, D. W. (2007). *How computer games help children learn.* New York, NY: Palgrave.

Shute, V. J., & Ventura, M. (2013). *Measuring and supporting learning in games: Stealth assessment.* Cambridge, MA: The MIT Press.

Tobias, S., & Fletcher, J. D. (2011). *Computer games and instruction.* Charlotte, NC: Information Age.

VanLehn, K. (2006). The behavior of tutoring systems. *International Journal of Artificial Intelligence in Education, 16*, 227–265.

Wouters, P., van Nimwegen, C., van Oostendorp, H., & van der Spek, E. D. (2013). A meta-analysis of the cognitive and motivational effects of serious games. *Journal of Educational Psychology, 105*, 249–265.

Index

P. Wouters, H. van Oostendorp (eds.), *Instructional Techniques to Facilitate Learning and Motivation of Serious Games*, Advances in Game-Based Learning, DOI 10.1007/978-3-319-39298-1

H

I

T

U